Groundwater Politics

Studies in Environmental Anthropology and Ethnobiology
General Editor: **Roy Ellen**, FBA
Professor of Anthropology, University of Kent at Canterbury
Interest in environmental anthropology has grown steadily in recent years, reflecting national and international concern about the environment and developing research priorities. This major new international series, which continues a series first published by Harwood and Routledge, is a vehicle for publishing up-to-date monographs and edited works on particular issues, themes, places, or peoples that focus on the interrelationship between society, culture, and environment. Relevant areas include human ecology, the perception and representation of the environment, ethno-ecological knowledge, the human dimension of biodiversity conservation and the ethnography of environmental problems. While the underlying ethos of the series will be anthropological, the approach is interdisciplinary.

Recent titles:

Volume 35
Groundwater Politics
Advanced Extractivism and Slow Resistance
Sally Babidge

Volume 34
Representations of "Japanese Nature"
A Historical Overview
Emiko Ohnuki-Tierney

Volume 33
Desert Entanglements
The Making of the Badiya by Sahrawi Refugees of Western Sahara
Gabriele Volpato

Volume 32
Fig Trees and Humans
Ficus *Ecology and Mutualisms across Cultures*
Yildiz Aumeeruddy-Thomas and Martine Hossaert-McKey

Volume 31
Sentient Ecologies
Xenophobic Imaginaries of Landscape
Edited by Alexandra Coțofană and Hikmet Kuran

Volume 30
Living on a Time Bomb
Local Negotiations of Oil Extraction in a Mexican Community
Svenja Schöneich

Volume 29
Grazing Communities
Pastoralism on the Move and Biocultural Heritage Frictions
Edited by Letizia Bindi

Volume 28
Delta Life
Exploring Dynamic Environments Where Rivers Meet the Sea
Edited by Franz Krause and Mark Harris

Volume 27
Nature Wars
Essays around a Contested Concept
Roy Ellen

Volume 26
Ecological Nostalgias
Memory, Affect, and Creativity in Times of Ecological Upheavals
Edited by Olivia Angé and David Berliner

For a full volume listing, please see the series page on our website: https://www.berghahnbooks.com/series/environmental-anthropology-and-ethnobiology

Groundwater Politics

Advanced Extractivism and Slow Resistance

Sally Babidge

berghahn
NEW YORK • OXFORD
www.berghahnbooks.com

First published in 2025 by
Berghahn Books
www.berghahnbooks.com

Library of Congress Cataloging-in-Publication Data
Names: Babidge, Sally, author.
Title: Groundwater politics : advanced extractivism and slow resistance / Sally Babidge.
Description: New York : Berghahn Books, 2025. | Series: Studies in environmental anthropology and ethnobiology; Volume 34 | Includes bibliographical references and index.
Identifiers: LCCN 2024046826 (print) | LCCN 2024046827 (ebook) | ISBN 9781805398820 (hardback) | ISBN 9781805398837 (epub) | ISBN 9781805398844 (adobe pdf)
Subjects: LCSH: Groundwater--Political aspects. | Aquifers. | Mineral industries--Political aspects. | Mineral industries--Social aspects.
Classification: LCC GB1001.7 .B32 2025 (print) | LCC GB1001.7 (ebook) | DDC 333.91/04--dc23/eng/20250111
LC record available at https://lccn.loc.gov/2024046826
LC ebook record available at https://lccn.loc.gov/2024046827

British Library Cataloguing in Publication Data
A catalogue record for this book is available from the British Library.

ISBN 978-1-80539-882-0 hardback
ISBN 978-1-80539-883-7 epub
ISBN 978-1-80539-884-4 web pdf

https://doi.org/10.3167/9781805398820

For Nancy

CONTENTS

Illustrations

Figures

Figures are © Sally Babidge unless otherwise noted.

Tables

Preface

Groundwater abstraction by the mining industry threatens to exhaust ecologies in many parts of the world. The 'critical minerals' rush, involving a multiplying frontier of extraction aspiring to feed the so-called green economy, pushes further into the possibility of ecological depletion. In the Salar de Atacama, Chile, part of a region renowned for copper mining and an area said to hold around a third of the planet's lithium reserves, locals, researchers and activists are concerned with how the new boom will affect the Salar's ecologies given the vast quantities of groundwater extracted to isolate and process these minerals.

For Atacameño-Likanantay (Indigenous) peoples, the lithium and copper mining industries have histories, differentiated, but each entwined with matters of groundwater. This book considers the ways that Indigenous peoples' engagements in the industrial past and present represent their efforts to actively generate a (good) future in the face of multiple potential harms from extractivism. It seeks to intervene in the global story of mining impact by focusing on the emplaced histories of transnational mining companies in their engagements with the Indigenous people from whose territories they extract, and emphasizes the ecological and economic vulnerabilities laid bare by the capitalist propensity to grow, expand and exploit.

Part of the impulse for writing this book has been in reflecting on the political, economic and social relations of extraction as a global colonial order, one dominant in my native Australia as well as Chile. There are parallels across these nations that multiply and diverge under scrutiny. In the end, Australia is in the background of the book's narrative, with cameo appearances such as the blonde Australian CEO of BHP in an Atacameño story about the boss who signed a community-company agreement. Otherwise, as a work driven by ethnographic research located in the Antofagasta region, the global mining company role is played by Chilean actors, employees of Minera Escondida, Albermarle and SQM, and their various contractors. On the other hand, there are significant matters of connectivity between The University of Queensland (UQ, my employer) and the mining industry in Chile. UQ's 'Sustainable Minerals Institute' exports contract expertise and training in a range of professions, especially on environmental and social relations, running industry-researcher conferences on these matters (see chapter 6, 7) and has offices in Santiago. The connections between the extractive industry and scholarship in Australian universities have garnered some critique (Lyons and Richards 2013). However, a detailed analysis of corporate-state relations between the two countries and their institutions, focused on the extractive industries, has not to my knowledge been attempted. It would be a rich avenue for the study of late capitalist extractive 'development'.

The transnational connections have a personal compulsion: as a young teenager I lived in Santiago de Chile during the two final years of the Pinochet dictatorship, when my father was employed by an Australian entrepreneur to run a just-privatized Chilean public utility. The brief experiences I had then of people's creative potential to endure and resist in the face of brutal forms of authoritarian control and social and economic injustices oriented a lifelong interest in struggle. When I began the research for this book, an emerging social and protest movement demanding dignity and change marred conservative billionaire Piñera's first Presidential term (2010–2014). The movement was sporadic but growing at that time, with broad discontent at two decades of democratic governance, which had incrementally deepened the radical inequalities generated by neoliberal forms; the economic 'miracle' of the dictatorship. Events that punctuated the final stages of fieldwork for this book mark a radical return in Chile to a political agenda of social and economic justice, one inflected with discourses of environmental protection. It is shadowed by ongoing reliance on earths and waters as resources to produce economic wealth, and thus by Australian, Chilean and transnational mining capital, keen to continue extracting and consulting to the industry and regulators.

A childhood connection and other matters of privilege inspired and enabled the instigation of this research and bind my identity with the extractivist and colonial dynamics about which I write. Other matters of my subjectivity should be clear to the reader in the chapters that follow.

Acknowledgements

Research for this book has been ongoing for some years and I am indebted to many. My desire to continue the research that has led to this book, and my return visits to Chile owe much to the informed and critical conversations I enjoyed with Peine community members, sufficient interest among them in my questions and hedged analyses, and their alacrity in demanding accountability from outsiders. I gratefully acknowledge the members of the Atacameño Community of Peine, and the Junta de Vecinos, Peine, who agreed to allow me provisional permission to undertake research in their village and territories in the first instance (2009), and then formal authorisation from the leaders and assembly in subsequent visits, with each fieldwork period necessitating renewed authorisation from different sets of community leaders (elected roughly annually) and the changing assembly. Return visits were always partly because I have grown to not be able to imagine life without Peineño friends. One family made the welcoming step of offering a room to rent and they have cared for me since, for which I am forever in their debt.

In the first long period of fieldwork (July-December 2011) I sought research authorisation from community leaders of Socaire, Talabre and Camar, and in some cases was approved, and spoke to their elected leaders for which I am also very grateful for their time and interest. For a range of reasons, I continued to focus my efforts at understanding things in Peine, rather than pursue the comparison of other community views. I also acknowledge leaders and members of the Consejo de Pueblos Atacameños and the Municipality of San Pedro de Atacama, especially those who granted some time in interview at different periods.

I am grateful to staff of the companies Rockwood/Albermarle, Minera Escondida/BHP, and SQM who agreed to speak to me for interviews, and employees of the smaller contracting companies that work on the Salar. Staff of the DGA in Antofagasta provided time and interviews for which I am thankful.

The research and writing has been made possible thanks to grants from the University of Queensland (including an International collaborative research travel grant 2008, Global Strategy and Partnerships grant 2017–18), institutional support for two semesters of Study Leave. An Australia Research Council Discovery Project grant (DP1094069 2010–2013) funded the early years of the project. A 6-month period of fellowship with UQ Institute of the Advanced Studies in the Humanities in 2016, and Visiting Professor fellowship at the Centre for Intercultural and Indigenous Research (CIIR, 2017) (CONICYT/ FONDAP/15110006) were financially and temporally indispensable for intensive periods of research, writing and community engagement.

Those who do not appear in these pages can be thanked by name. Rene Huerta Quinsacara in Antofagasta, and Cristina Garrido in San Pedro de Atacama, were my first hosts and friends in the north, invited me to stay in their homes and have been critical friends and teachers. Thanks for friendship in remote places also to Paula Vargas, and in Santiago, John Paul Chamberlain and Rodrigo Molina sustained me with laughs, wine and food.

Colleagues with anthropological and other expertise in the region steered early and subsequent work. Deep gratitude is due to those who facilitated visitor status at different times at the Universidad Católica del Norte, Institute and Museum (IIAM) in San Pedro de Atacama. Special thanks to Saúl Cervantes, for expert knowledge and assistance in accessing material at the IIAM/UCN library. Being a visitor at CIIR in Santiago, at the Pontífica Universidad Católica de Chile, in the excellent Anthropology Department, and at the Anthropology Department of the University of British Columbia, both in 2017, were deeply influential on early drafts of the book and colleagues at each institution influenced my thinking, writing and manner of ethnographic relating.

Thanks to archival collections: Museo de Arte Precolombino (Santiago), Museo Nacional de Historia Natural (Santiago) and National Library (and archives) of Buenos Aires, Argentina.

Thanks to colleagues at the CSRM (SMI), UQ, who agreed to waive the corporate fees to attend their conferences and at times have invited me to attend industry-linked events that were otherwise not on my radar.

Early entries in my digital bibliographic library and interview transcription files are haunted by the particularly effective research assistance of Joanna Horton, Yuri Yudelevich, and Kevin Porter. The Ethnography Reading Group at UQ has accompanied and challenged my thinking while writing. Discussions among advanced students of the undergraduate course 'Anthropological Futures' since 2020 have accompanied redrafting of this book over the last four years. Current and previous PhD students at UQ, Charmaine Jones, Kaylen Jorgensen, Carla Meurk, Isabel Cane, Carlos Rivera Santana, Alex Gearin, Gillian Paxton, Maria Itati Dolhare, Madeleine Belfrage, Tribute Mboweni, Francisco Tiapa, Terry Morgan, Nigel Wight, Sara Mejia, Pablo Aranda, Nathalia Dothling-Reis, Sarah Thomson, Diana Romano, and Ricardo Labra have each influenced thinking about this book project through our conversations about their work, or have worked with me as colleagues on aspects of the research that appears in some guise or another in these pages.

For (variously) reading drafts, making comments on or for conversations about different chapters, and for important intellectual nourishment, I thank colleagues and critical friends: Lillian Geddes, Naomi Radunski, Patricia Ayala, Alonso Barros, Paola Bolados, Anita Carrasco, Milka Castro, Francisco Rivera, Hans Gunderman, Anita Lemus, Dante Choque, Fernanda Kalazich, Manuel Prieto, Valentina Figueroa, Leslie Robertson, Terre Satterfield, Shaylih Muehlmann, Juan Francisco Salazar, Sol Rojas-Lizana, Roberto Esposito, Erin Fitz-Henry, Monica Minnegal, Casey Walsh, Jeremy Walker, Marina Weinberg, Cristobal Bonelli, Piero DiGiminiani, Giovanna Bacchidu, Marcelo Gonzalez Galvez, Pedro Mege, Sonia Roitman, Anna Pertierra, Celmara Pocock, Nick Bainton, Deanna Kemp, Ciaran O'Faircheallaigh, Allison Fish, Brad Sherman, Susannah Chapman, Veronica Strang, Gerhard Hoffstaedter, Kim de Rijke, Peter Harrison, Anna Johnson, Sarah Holcombe, Kristen Lyons, Javiera Barandiarán, Tom Özden-Schilling, Timothy Neale, Tom Powell Davies, Linda Connor, Gillian Cowlishaw and Ute Eickelkamp. Nancy Williams read every early draft a few times over, and my expectation of her pencil mark when I type is testament to her influence. Three anonymous reviewers of the MS were generous to read it in detail and make excellent suggestions. Any errors land at my doorstep.

Thanks also to Merilyn, Mark (dec.) and Kate, who heard all about the long process of writing this book and nonetheless expressed interest and support. To David for surprising tangential commentary and outstanding company.

Parts of Chapter 2, 4 and 7 have appeared in previous publications, but the data and analysis have been updated for the present text.

ABBREVIATIONS

APR	Agua Potable Rural – Rural Potable Water organization
BHP	BHP (mining company, majority owner and operator, Escondida mine)
C169	Convention 169 of the International Labor Organization
CAR	Comité de Agua Rural – Rural Water Committee
CDE	Comisión de Defensa del Estado – Council for the Defence of the State
CLP	Chilean Pesos
CODELCO	Corporación Nacional de Cobre – the National Copper Corporation
CONADI	Corporación Nacional de Desarrollo Indígena – National Indigenous Development Corporation
CORFO	Corporación de Fomento de la Producción – Chilean Economic Development agency
CPA	Consejo de Pueblos Atacameños – Council of Atacameño Peoples
CSR	Corporate Social Responsibility
DGA	Dirección General de Aguas – Chilean Water Authority
DIA	Declaración de Impacto Ambiental – Declaration of Environmental Impact
DOH	Departamento de Obras Hídricas – Department of Water Works
EIA	Estudio de Impacto Ambiental – Environmental Impact Study
Escondida	Also MEL, Minera Escondida Limitada (mining company)
FME	Fundación Minera Escondida (philanthropic foundation)
ILO	International Labor Organization
JV	Junta de Vecinos – Neighbourhood Council
L/s	Litres per second
MEA	Millenium Ecosystem Assessment
MNT	Monturaqui-Negrillar-Tilopozo

PAT	Plan de Alerta Temprana – Early Warning Plan
SAG	Servicio Agrícola y Ganadero – Agriculture and Livestock (Government) Service
SEA	Servicio de Evaluación Ambiental – National Environment Evaluation Agency
SEIA	Sistema de Evaluación de Impacto Ambiental – National Environmental Impact Assessment System
SCL	Sociedad Chilena de Litio (mining company)
SQM	SQM Salar (mining company)
SMA	Superintendencia del Medio Ambiente – Superintendant of the Environment
USD	United States Dollars
Zaldívar	Compañía Minera Zaldívar (mining company)

Introduction

What Surfaces

Unlike other superficial water, Don Geraldo explained as we sat in the shade at Tilopozo, this spring formed a permanent pool, because '*afloran todas las aguas*', the waters that surface here on the southern edges of the Atacama salt flat flow underground from all the Andean Mountain peaks.[1] I asked, 'have you seen Tilopozo change since the beginning of mining'? He paused before replying, 'just by looking it is difficult to say . . . but you can see when water is coming down and when it does not'. The landscape surrounding Tilopozo seems stark to an unpractised eye; there are few trees or other vegetation, except in village oases. Rain and snow melt from the peaks seeps into the earth, becomes groundwater and forms springs in the steppes and foothills and fresh and saltwater pools on the salt flats. Freshwater surfacing at Tilopozo evaporates through grey and green vegetation, forming a briny meadow.

What surfaces in the pool and nourishes the pasture at Tilopozo includes water that began as snowfall in the peaks of the Andes range to the south and west, and if there is less snow and rain, or if the subterranean flows are interrupted, then there will be less water on the edges of the Salar de Atacama. But unlike Gerardo, whose lifetime in the southern Salar, and of travel across the Andes for trade or into the highland pastures with herds, has depended on knowing the location and ensuring the quality of springs and other water sources *en route*, I could not 'see' the waters to which he refers.

The Salar de Atacama, in the interior of Chile's Region II, Antofagasta, is part of a region synonymous with mining and its twinned potential for economic wealth and ecological exhaustion. The mountains are threaded with rich veins of copper, proven by ancient mines worked by Atacameño-Likanantay peoples, and mega-voids of contemporary industry. A booming interest in lithium salts is driving the extraction

Notes for this chapter begin on page 26.

of new mineral assets that are suspended in the Salar's waters. But the Salar de Atacama has come late to the wealth and violence of the globalizing extractivist story. Partly because of its scant surface freshwater, the growth of mega mines began only at the end of the twentieth century, when groundwater pumping technology advanced sufficiently to allow for the lubrication of copper and lithium processing need. The industry has swelled again in the early twenty-first century, as the Salar's reserves of lithium have been the subject of a mining boom for 'energy transition' minerals.

Mining for copper and lithium involves extraction of vast quantities of groundwater. However, the surface changes from groundwater pumping evolve slowly; uncertainty is more common than clear evidence of 'mining impact', ecological harm. A mode of ethnographic attention, particular to anthropologists, is to take note of the ways in which 'facts falter' in a 'world beset by uncertainty' (Stevenson 2014). However, Atacameño people like Gerardo insist that ecological harm must have occurred over the many decades of copper mining company pumping from aquifers and have become increasingly concerned about the effects of the lithium industry's water extraction. Under the conditions of Chile's neoliberal political economy, transnational companies are responsible for monitoring subsurface and surface environments subject to their activities, and until recently, there has been an absence of independent or state regulator assessments. This leaves extractive interests largely free to define the facts, the material capacities of groundwater, and wrangle all the data.

That water, as material, is seldom still and resists containment, does not prevent extractive assertions about its predictability. A mining company-employed geohydrologist characterized water as 'very intelligent . . . always want[ing] to go the easiest way.' If water is animated for company science, then it is unidirectional and restricted to water's imputed desire for ease, which in turn allows water to be represented in company reporting as entirely predictable.[2] Against these assertions, Atacameño peoples have struggled to prove mining impact on surface ecologies such as Tilopozo, despite the quantities of groundwater pumped from beneath it.

This book examines a recent history of relations among extractivist and Indigenous ecologies and humans of the Salar de Atacama, their contests and divisions, as well as agreements and intimacies. As an ecological actor, groundwater relates to air as it gurgles upward to form a spring and nourishes plant and animal life creating features in the landscape. The diverse human relations of the Salar de Atacama are nourished by water, especially groundwater. Their relations are evident in the ways that fresh and saltwater's scarce surface appearances in the Salar de Atacama partly

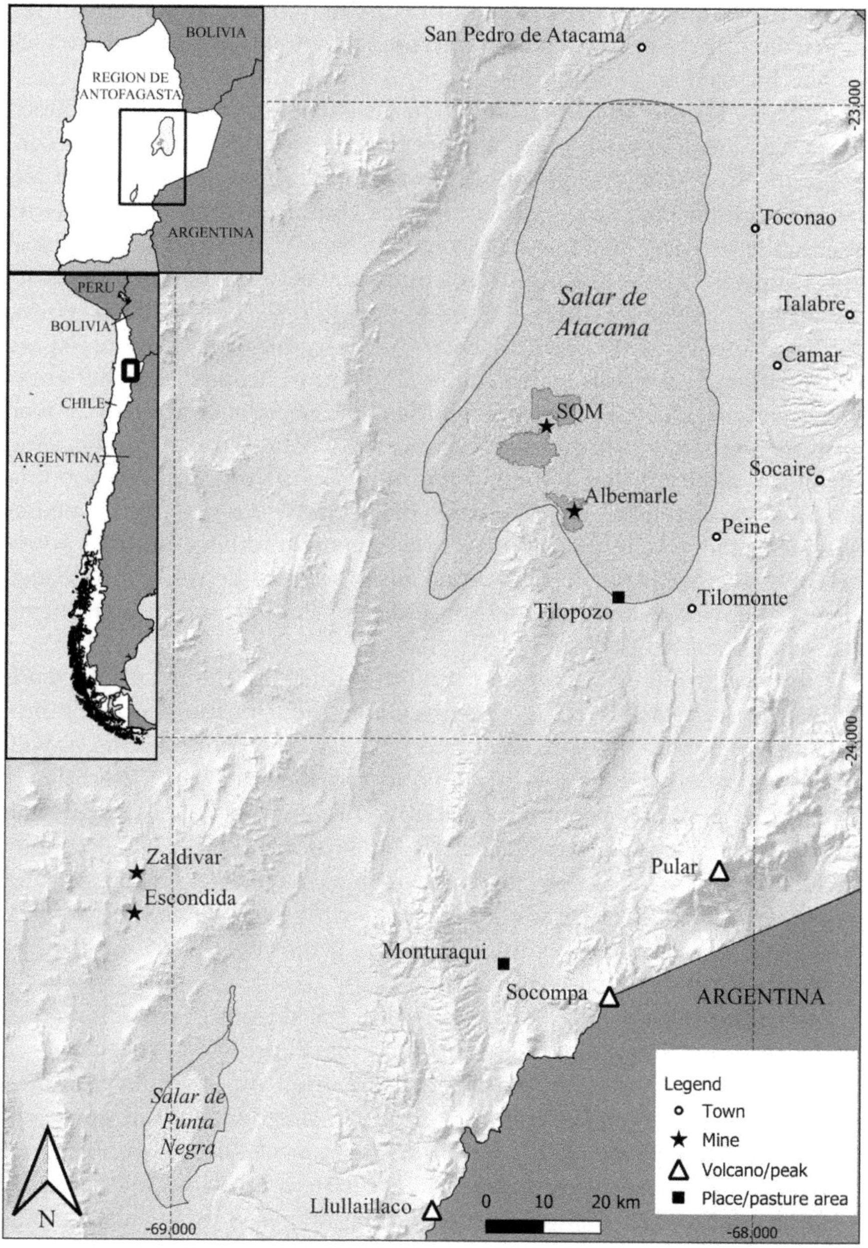

Figure 0.1. Map of the Salar de Atacama. © Sally Babidge and Jasmine Boyton.

define the parameters of Atacameño-Likanantay (Indigenous) territories, which in turn affect which mining company must engage with which named Indigenous community.

Water's absence on the surface is unmistakeable in the arid surrounds of the Salar. In contrast, maps and models of groundwater show it congregating from many directions, high altitude to low, to pool in subterranean aquifers, seeping in, moulding the channels where it runs along, eventually arriving at the surface to combine with dirt and salt in places like Tilopozo. This is groundwater's material activity – where it flows to, bubbles up, or stays hidden – with its potential for geo-political effects. As geographers of the 'hydro-social' have argued, sociopolitical space is constituted through water-society relations, cycles and networks (Swyngedouw 2009; Linton and Budds 2014; Boelens et al. 2016). Given its more intangible properties, groundwater may have social, political and ecological relations that are less constitutive, emplaced, or even good. In the Salar de Atacama, the properties of groundwater and the capacities of water generally, in its contingent relations with others, nourish political contest, a particular form of relation, among and between Indigenous groups themselves and the interests and practices of transnational corporate extractors and the state.

With the growing copper and lithium industries, the Salar's groundwater seems to have at once expanded in its capaciousness to accommodate growing volumes of extraction recorded in water concessions, as well as having become more diminished and threatened. The indeterminacy of groundwater presents both a scientific problem as well as a question about governability (Birkenholtz 2015; Molle and Closas 2019; Gleeson et al. 2020). As human knowledge of and reliance on aquifers for freshwater has grown, so have concerns about groundwater pollution and depletion. UNESCO, for example, announced a year of groundwater in 2022, and the challenge to 'make the invisible visible'.

Institutional knowledge of aquifers and the science of modelling and measuring aquifer systems is an anthropological problem in that making the subterranean world 'visible' is a matter for how diverse human knowledge works, and how environmental objects are defined and governed (Ballestero 2019; Walsh 2022). Groundwater's 'invisibility', an object not available to the naked eye, means that access to it is always mediated; its location, activity, quality, and so on can be known only by surface outflows such as springs, or by using technologies that perforate the surface for the collection of data used in the construction of hydrological models. The earth's surface and modelling technologies are the narrators of the story of human-groundwater relations. Quantities of groundwater are visible in government registers and environmental management models as abstract volumes, and waters that show up in this way connect diverse sets of people to one another, and to projections of social and ecological futures. Groundwater may be characterised in terms of aporia (Bessire 2022), never lacking in definition, projection or expressions of concern, but without any consistency and apparent truth value among these. The material fact of groundwater's surface invisibility may be less a problem than the political economic fact of capitalist propensity to exploit it.

Surface ecologies provide few environmental clues about harm from extraction until many years after the damage is done, especially in the case of 'relic' sources such as those thought to sit beneath the Salar de Atacama (Moran et al. 2022). The lengths of snaking black pipe, large holding tanks, leaking outlets and discarded mining equipment are contrary signs of industrial disaffect. The visual cues of extractive effects have become more apparent with the lithium industry expanding; groundwater concessions have multiplied on government registers, the horizon is increasingly hazy, old pastures apparently crunchier and village water supplies under greater demand with local population growth and in-migration of people attracted to wage labour and other economic opportunities.

The Salar, having become renowned as a promising and rich mining frontier, is experiencing a shift in public concern for its ecology. International attention has been drawn to matters such as the ironies of the 'green economy'; the role of mining lithium for energy transition, the exponential growth of that industry, and embedding cumulative harms from the long-standing effects of the copper industry. In the mounting critical literature focused on the lithium boom, Atacameño-Likanantay peoples are represented as 'suffering the impacts' of mining and the combination of these industrial and representational dynamics has intensified talk among some locals, as well as other commentators, of an ecological crisis. Our attention to the injustices of dispossession wrought by extractive capital accumulation demands a just response. However, a focus on crisis may distract from elemental and structural questions that can point us in the direction of what counts as 'normal' (Roitman 2013). Crises generated by extractivist exploitation are worth attending to. This book argues for an equal attention to the things that happen below the threshold of eventfulness, below the surface and what passes in plain sight; the mundane or ordinary wearing-out, the slow deterioration of relations, and incremental acts by which relations may be denied.

Detailing recent histories of extractive activity in the Salar de Atacama, this book frames the effects of mining in relational, material (ecological and economic) and political terms. Extractive capital not only creates ecological harms and enhanced social inequality (Sassen 2014, Nixon 2011), but it also produces significant wealth for companies and states and promulgates 'community development' (capital investment in health, education, housing, infrastructure and local business). Studies undertaken by and about the industry that claim to make transparent these countervailing effects, including but not limited by 'impact' studies, drive intensifications, accretions and disjuncture in scientific, Indigenous and local environmental knowledge. The ethnographic work for this study began in 2009, well before the current sense of overwhelming and rolling planetary crises were somewhat normalized, at the cusp of new environmental regulations in Chile, and with the expansion of the Salar de Atacama's lithium mines on the near horizon. While aspects of this study trace how environmental and social conflict associated with extractive industries has surfaced over the last decade or more and is attentive to the urgency of responding well to potential and actual ecological harm, it also argues for responsiveness to the slower moving forces in the powerful effects of extractive capitalism.

A Material Politics

In Atacameño-Likanantay localities, people pipe spring water into domestic taps, use it to run commercial kitchens and other small businesses, and maintain irrigation infrastructure so that the waters nourish their maize, potatoes and other crops grown in agricultural terraces. In 2009, members of the Atacameño Community of Peine community were filmed kneeling at the foot of Llullaillaco peak. The *cantal* (ceremonial leader) is recorded saying,

> Pachamama, Saint Earth (*Santa Tierra*), with all respect, Pachamama, we have come to this place, face to face with the cerro Llullaillaco. I ask of you, as always that our water never diminishes. Because you are the one amongst the others, and all the cerros have water and have springs. And these waters are for the animals that you have, you are owners of the herds that are here in the countryside. And we have always asked that it rains, that the rain makes the pasture for animals, that the country is maintained green. Pachamama, Saint Earth, receive this small convido, Saint Cerro, Llullaillaco, so powerful that you are, so far away and we have come to be close, but that is how it is when one wants to be close to another . . . (in Gedda 2010, 27:40 mins)

Those depicted in this scene of the film were testifying to the Atacameño Community of Peine's territorial claims. Over the years previous, community leaders in Peine and neighbouring Atacameño communities had engaged anthropologists and lawyers to help document the full extent of their ancestral relations to territory. Communal title to village territory and water was secured, including the named springs that nourish Peine's agricultural lands, and in Tilomonte, the locality that holds fields and springs of their second oasis settlement some fifteen kilometres south. However, the familial histories of herding and ecological knowledge of the Salar, foothills and highlands, were not fully counted in the legal rights recognized by the state. Excluded from full recognition of territorial rights are many watered places in the memorial cartographies that trace paths across contemporary national borders with Argentina and down to the refuges and waypoints that lead to lowland oases and the Salar.[3]

Historically, projects responding to Indigenous people's voices regarding the harm done by extractive industries have focused on such recognition of Indigenous *rights* to water and territory (Yáñez and Molina 2008). But granting Indigenous rights to surface water and territory in Chile has not resulted in any radical reordering of the inequities of water access for people, nor, it seems, an overall reduction in extractive harm to ecologies. This is not simply because those rights have never fully been actualized by an ineffective state regulator. In the Salar de Atacama, as elsewhere, a legal and political economic distinction between groundwater resources available for capital exploitation and rights to surface water for the support and protection of community and ecological life is a convenient fiction for ongoing extractive activity.

Of course, the surface and subsurface are connected. However, the legal distinction maintains one separate from the other, containing groundwater in resource ideologies, as geophysical material, as volumes of owned rights to the groundwater, and as data in models that enable a calculus of extraction. This material politics, which is also a knowledge politics (Barry 2013), is critically important for arguments made by industry relating to the sustainability of their extractive activities. This book proposes that a material politics, in particular, an attentiveness to the diversity and negotiations in people's relations among one another in relation to earths, water and territory, informs how to address economic injustices and ecological harms. My usage (explored in detail in Chapter 1), is in the sense of both an older materialist analysis concerned with the political and moral economic dynamics of global extractive capitalism generating and sustaining inequalities and injustice, as well as a new materialist concern with the contingent capacities of non-human beings and things (for example, ecological things like peaks) in their relations with humans.

In the territories the Peine community considers its own, mining projects were initiated in the 1980s. Two copper miners extract groundwater and operate large open cut mines in this area. Minera Escondida Limitada (Escondida), a mega copper mine majority owned and operated by BHP, with partners Rio Tinto and Japanese consortium JECO, and Compañía

Minera Zaldívar (Zaldívar), operated by Antofagasta Minerals, which at the time of writing was a joint venture between the Chilean company Antofagasta and transnational Barrick Gold. Both mine copper at high altitude around 100 kilometres to the south of the Salar de Atacama. These operations demand of their workers a high level of technical training and they have employed only a handful of Atacameño people over the many years of their operations. Escondida and Zaldívar control the highest volumes of groundwater extraction rights of all corporate interests, pumping from an aquifer on the south of the Salar de Atacama.

In 1997, the Peine Atacameño Community and the copper miner Minera Escondida signed one of the first legal agreements for 'development benefits' in Chile between a mining company and an Indigenous community. With this, and through its not-for-profit social projects arm, the Minera Escondida Foundation (FME), Escondida would deliver development benefits to Peine partly due to, but not in explicit (contractual) recognition of, company groundwater extraction in Peine's claimed Indigenous territory. Until very recently, Zaldívar has had little to do with the community.

Lithium mining in the Salar de Atacama is undertaken by transnational corporate operator Albemarle (USA-based) and the majority Chilean-owned 'Sociedad Química y Minera de Chile', or SQM. These operate on the south and western edges of the salt flat, and are effectively brine (salt water) extractors, pumping groundwater viscous with minerals into a series of shallow ponds where evaporation leaves key elements behind. These operations, and the large number of contracting companies that service them, have always employed local men from Peine and their neighbouring villages.

In 2009, when I began to ask members of the Peine Atacameño Community about their relations with Escondida and other mining companies, those I spoke with emphasized the threat to their territories from copper mining company groundwater pumping. Reports on and questions about the impacts of groundwater pumping by Escondida and Zaldívar dominated the community meetings I attended over the course of more than a decade of visits to their village and the Salar, as did criticisms concerning the negligible effects of development benefits from their agreement with Escondida and ongoing problems with the terms of the agreement and company proposals. However, the material politics and relations with lithium extractors are somewhat different.

Unlike the open-cut scars of the copper mines concealed in the mountains to the south of the Salar, exhausting invisible groundwater from beneath it, the lithium mines of Albemarle and SQM can be seen from most points in the village of Peine. Each year, over the decade or more of visits to conduct fieldwork in the village, the smudged white lines on the western horizon have become more evident during the day, the layer of haze atop thicker, and the twinkling row of lights at night longer and brighter. The parade of headlights in the evening heralds the return of lithium mine workers and contractors to company camps on the edge of Peine village and to accommodation options expanded by enterprising local households. In recent years, saltwater pumping by Albermarle and SQM Salar has grown, and the twinned ecological and economic effect of these mines is more intensive in terms of the relations themselves and uncertain and complex in relation to the elements of the relations at play.

At the time of writing, the Atacameño Community of Peine has (as have other Atacameño-Likanantay communities of the Salar) entered into agreements with each of the brine extracting companies in their efforts to harness some of the economic benefits being extracted and use agreement processes to push the state for regulation and mitigation of mining's ecological harm.

Mining companies issue public reports on sustainability measures, setting out company data on territorial and subterranean water extraction within legal concessions. Nonetheless, Peine community leaders actively seek to visit mines and locations of extraction infrastructure in groups and accompanied by company social relations employees (who facilitated such visits) in attempts to see the effects of industrial activity for themselves. During one trip in which I participated, standing next to a large pump, a woman very familiar with company reporting wondered aloud, 'how much water is *really* being taken from here?' Questions such as this were asked with an awareness that there was no satisfactory reply. A material politics is a useful analytic for examining questions about the activities of extractive industries and contestation arising from their presence, infrastructure and claims in corporate documents provided in the name of 'transparency' (Barry 2013).

Surface and subsurface ecologies and their relations are either muted or enhanced during extractive activity and development projects, which for their effectiveness, rely on ostensibly public but remote-from-view relationships between mining companies and Indigenous communities. Water imagined and materially represented in management registries, unseen but suspected in infrastructure gauges, and supposedly flowing through pipelines, are instances of active and contingent ecological relations. These waters may have a negative relation to communities (out of sight, out of their hands, polluted or of low quality), and potentially also to company actors (leaking pipes, the excess from production or the cost of concessions). Relations may be of a negative kind, but attention to them highlights their persistence, intimacies and generativity (Strathern 2020). The recent history of Atacameño-led, extractive industry-funded development in the Salar is marked by material uncertainties about mining's productive potential for economic wealth and of disconnecting relations between groundwater and human and other life. These uncertainties shape the contours of mining company-Indigenous community relations. A combination of economic and geophysical factors helps materialize the potential for life to flourish and falter under conditions of advanced extractivism. This book explores these questions.

Advanced Extractivism

As mining technology enabled extensions downwards into the earth and its aquifers, the political economic and legal conditions of late twentieth century neoliberal capitalism empowered multinational companies to grow mines. In concert, an explicit discourse about good industrial behaviour developed among leading mining company actors, who began asserting that a 'socially responsible industry' promising to fill the gaps in deficiencies of state regulatory regimes and operate in ways that are 'sustainable'. Each of these conditions – vertical technologies, an extractivist political economy, and corporate social responsibility, or 'ethical' capitalism – have intersected to intensify relations between industry and communities local to extractive activities. This mix of technological, material and moral factors is what I mean by 'advanced extractivism'.

In order to set out the concept, I weave together a history of Chile's changing neoliberal political economy with the critical literature led by Latin American scholars referring to 'extractivism'. I provide an account of Chilean political economic change that is relevant to the legal and environmental histories of extraction and state policies that have advanced industry and capital relations of these kinds, particularly in the north of the nation. This provides the basis for arguing for the three principal conditions of advanced extractivism.

Chile's Extractivism: Mining and Neoliberalism

Extractivism refers to a political and economic project that promotes the exploitation of earths, minerals and nature principally in terms of them being resources for foreign investment and exports, and thereby to generate national economic growth, with the common effect of large-scale degradation of ecosystems and expansion of exports to international markets. The history of extractivism in Chile is rich. In the broader Antofagasta region in which this book is set, the state-run Chuquicamata copper mine adjacent to Calama has been the focus of studies considering the nation's economic reliance on copper and entanglement with north American and 'global' mining.[4] 'Chuqui' may be among the biggest, but it is far from the only copper mine in the region, nor is copper the only extractivist concern that has created the north as a mining region populated by miners (Babidge and Belfrage 2017).

Chile is also infamous for its extractive history in nitrates, that, like the copper industry, produced significant wealth for some in the nation, and a site of violence and oppression for workers and their families (Frazier 2007). The critique embedded in the term 'extractivism' points to anthropocentric relationships to ecology-as-resource (earth as inherently exploitable), to the colonial legacy of exploitation dating to Spanish conquest (Assadourian 1982; Galeano 1971), and to asymmetric power structures of globalizing mining capitalism with its effects of violence and dispossession (Gómez-Barris 2017; Sassen 2014). In addition to destructive social and ecological practices, extractivism implies a logic of political and economic practice, a way of thinking and being (Willow 2018). Under neoliberal capitalism of the late twentieth century, extractivism's frontiers expanded, as did capital exploitation of territories and associated waters. Neoliberal governments in Latin America, especially in Chile (Folchi 2009), have been important contributors to the advanced extractivist dynamic.

As Harvey (2005) outlined, neoliberalism has been an extraordinarily successful political economic system in terms of spread across diverse jurisdictions. The Pinochet regime's neoliberalism, cemented in the nation's constitution and expanded by the democratic coalitions that followed, makes Chile's history of extractivism an exceptional case.[5] Anthropologists and others have shown that everywhere extant forms of state structure and corporate articulations, as well as cultural personhood (e.g. Ferguson 2010; Rajak 2011b; Shever 2013) only ever have, in Tsing's (2005) words, 'makeshift links' to a global form. This is an argument about documenting 'actually existing' neoliberalisms (see Wacquant 2012).

In the 1970s, Chile was a laboratory for the north American economic theorists and the International Monetary Fund (IMF) to test the potential of 'liberating' markets through expansive privatization. Neoliberalism was imposed by Augusto Pinochet's dictatorship (1974–1989) enforcing the replacement of the world's first democratically elected socialist government with violence and through the eradication of workers unions, economic policies that privatized utilities and goods according to the dictates of the 'Chicago Boys', and constitutional change that protected corporate freedoms. Mining, especially copper, grew under both private investment and state-run business, with the regime retaining hold of Chuquicamata and three other mines' operations functioning as the 'Corporación Nacional del Cobre' (CODELCO).[6]

Under the Water Code (1981), water was decreed to be 'freely bought, sold, mortgaged and transferred' (Bauer 1997: 641). The broad-reaching neoliberal experiment in Chile, in which the Water Code is significant, was hailed by some as an exemplary case for the role of the private sector in allocating and making best use of resources for national development (Briscoe et al. 1998; see also Senate Standing Committee on Foreign Affairs Defence and Trade 1992). Water law was central to the formation and economic effectiveness of neoliberalism in Chile and was critically paired with the reproduction and deepening of social inequalities of access and use (Tecklin et al. 2011; Barton et al. 2012; Budds 2012).

Early twenty-first century neoliberalism's different expressions were emplaced across Latin America and elsewhere through 'structural adjustments' imposed by the IMF (Bauer 1998, Brown 2019). That post-Dictatorship and liberal democratic (*Concertación* coalition) governments in Chile perpetuated and deepened neoliberal forms introduced under the terms of Pinochet's constitution, with its privatization of education, health, welfare and water, is a well-researched historical and political economic wonder (Moulian 1997; Paley 2001; Winn 2004; Alexander 2009).

The progressive legislation introduced by the first democratic regimes of the 1990s, including the Indigenous Act (1993) and an Environmental Act (1992), moved to protect surface water sources in the interests of human and other life. The legislation conformed to liberal economic norms and expectations of global civil society. Provision was given to the recognition of waters as owned and inalienable in the name of titled indigenous communities where these waters appear on the surface (Ministerio de Justicia República de Chile 2008; Latta and Cid Aguayo 2012). The Chilean Water Authority (DGA) and the National Indigenous Development Corporation (CONADI, established under the Indigenous Act) signed an agreement to 'protect, constitute and re-establish' water rights for Indigenous peoples, with a special focus on the northern desert regions (Castro-Lucic 2002: 198). Indigenous communities were able to own only the resources directly related to domestic and agricultural (that is, productive) activity (see MacPherson 2020), with water sources in the expansive territories associated with customary pastoral movement more difficult to secure (Barros 1997).

The ecological harm done by extractive industries in the Salar de Atacama (and elsewhere) through groundwater extraction is held up against the weakness of law in relation to the recognition of Indigenous rights to water and territory (Yáñez and Molina 2011). However, Chile's neoliberalism defends and consolidates rights and law, as others found in relation to Human Rights discourses, in which 'constitutional restraints and law [are considered] necessary to preserve the individual freedom that only a competitive market could ensure' (Whyte 2019: 3; cf. Goodale 2016). The granting and exercise of Indigenous and human rights under progressive legislation of democratic governments, has not resulted in any radical reordering of the inequities of water access for people, nor, it seems, an overall reduction in extractive harm to ecologies. Such avenues have nonetheless seen incremental regulation of extractive harm.

President Michelle Bachelet's two terms of office (2006–2010 and 2014–2018) progressed the introduction of a range of reforms relevant to Indigenous peoples and environmental protection. In 2009, at the time that I was undertaking the first stages of ethnographic fieldwork that has led to writing this book, Bachelet's government ratified Convention 169 of the International Labor Organization (ILO, 'C169'), by which the state confirmed the rights of Indigenous Peoples to be consulted regarding any legal process, activity or decision that directly affected them. By 2010, the government revised and renewed the Environment Act to keep up with these changes, replacing older bureaucratic agencies and elevating the status of environmental concerns by creating a Ministry for the Environment, the Department of Environment (SMA) and Environmental Impact Assessment Service (SEIA) (Barandiarán 2018). The laws and new regulatory frameworks made special mention of extraction and mining projects, with specific decrees regarding the ways in which extractive projects would be assessed by the state before proceeding. These measures and others that brought Chile in line with global civil society and liberal expectations were also important in terms of their inclusion as one of the first of the Latin American nations integrated into the Organization for Economic Cooperation and Development (OECD).

None were able to dent the armour of Chilean neoliberal conditions of private property rights and market freedoms. Chile's water regime under neoliberal governance has been judged a spectacular failure for water equity and public access (Bauer 2004, 2013; Prieto et al. 2020), notable even by IMF commentators (Ostry et al. 2016). Moreover, expansive privatizations of public utilities, under the Pinochet civic-military dictatorship, have had long-term effects in regard to the capacity of government agencies in both scientific and administrative function (Barandiarán 2018), and prevented practical change as symbolized in apparent reform.

A growing student movement in the second decade of the twenty-first century demanded changes to education and welfare, but also to the widespread and entrenched inequalities of neoliberalism.[7] In October 2019, the so-called *Estadillo Social* (Social Outburst) saw millions of Chileans occupying the streets of the capital and the cities and towns of the nation. After a month of violent state attempts to bring order, Conservative President Sebastian Piñera announced the 'Agreement for Social Peace and a New Constitution' (15 November 2019).[8] In part, this involved commitment to a plebiscite to decide if Chile were to have a new constitution, and if so, who would write it. Significant street-level social mobilization continued, until conditions of the global COVID-19 pandemic drove protestors into their homes. The plebiscite was eventually held in late 2020. A massive majority of Chileans voted not only to replace the constitution but for the new constitution to be written by members of an elected citizen council, rather than the 'political class'.

In December 2021, Gabriel Boric, active as a student leader in mass protests of 2011, was elected President with 55.8% of the vote, and according to the official government website, the highest number of votes and largest electoral participation in Chile. He is quoted as saying before his election that 'if Chile was the cradle of neoliberalism, it'll also be its grave', and promised to represent a new environmental Left that would make social reforms without sacrificing the environment.

Taking office in March 2022, his government oversaw the process of elections for a Constitutional Convention. The constitution was rewritten by a convention of 155 elected members, seventeen of whom were First Nations peoples in reserved seats. Important aspects of the rewritten document focused on Indigenous rights exceeding any prior laws, including rights to culture, autonomy and self-government, on gender parity and on radical changes to water laws, including 'rights of nature'. The document was finalized in 2022, and in a September referendum, failed to receive popular approval. A second Constitution was written, heavily influenced by a committee membership of far-Right and conservative party members, but in late 2023, Chileans also voted that down (by a national majority of 55%, with voters in the San Pedro de Atacama sub-region coming out at around 70% against).

President Boric has nonetheless brought in changes significant to the material politics of the Salar de Atacama and its extractive industries. The first was to oversee the 'Water Code Amendment Bill' (Gobierno de Chile 2022b).[9] The reforms see the human right to water take prominence over private property rights, reduced water rights previously granted in perpetuity to thirty-year concessions, and enable the state to seize water rights in ecosystem or domestic interests. The Bill radically overturns the prior logic of the marketization of water in the old Water Code, but may have limitations, as Elizabeth MacPherson et al. (2023) argue, in terms of addressing water injustices in Indigenous territories.

Second, with the 'National Lithium Strategy' (Gobierno de Chile 2022a) made policy in 2023, the Boric government brought into being a massive investment in lithium resources, principally by aiming to be part of every lithium project in the nation. SQM signed an agreement with Chile's national copper company, CODELCO, to achieve 'nationalization' by 2031. Albemarle's concession in the Salar expires in 2043 and company spokespeople responded to the state's lithium policy by indicating that it was open to negotiating with the state. While the Chilean state undoubtably has an interest in lithium extraction, production and manufacture to fund the social agenda, the Lithium Policy proposes to modernize institutional forms, create a network of protected places in salt flats in Chile and increase national investment in research. The state's partial nationalization of lithium 'resources' through the Lithium Policy promises to safeguard ecosystems. It also sets out agreements (through CODELCO) to be made with the Atacameño communities of the Salar. These have so far been received with acquiescence by the industry and critical scepticism expressed in public statement by Indigenous community leaders. The latter have been clear about the ways in which the radical optimism of Boric's policies may not correspond with local visions of what is good for them and for the Salar.

Extractivism and 'Ethical' Capitalism

When Chile transitioned to democracy in the 1990s, other Latin American progressive and socialist governments had begun pairing regulation, Indigenous recognition and forms of benefit redistribution with extractive industry 'development'. This pairing has been called 'neo-extractivism' (Gudynas 2009, 2012; Svampa 2019) and refers to a critical account of the 'pink tide' or leftist governments that emerged in the latter part of the twentieth and early twenty-first century in Latin American states and which tended toward reliance on the extractive industries to fund broad sweeping social reform. Under neo-extractivist governments' progressive policies, including constitutional or other legislative recognition of Indigenous Peoples' rights, the Rights of Nature (in Bolivia and Ecuador), environmental protection and community consultation mechanisms, and the use of mining rents for social spending, old problems of ecological exploitation and social inequality remain (Acosta 2013).[10] As North and Grinspun (2016) argue, the social development focus of neo-extractivism has not addressed manifold inequalities in 'rural life' across the continent. Instead, land and resources continue to be transferred into the hands of local and national elites and global corporations.

Regulations regarding development and environment introduced by the Bachelet Government in 2010 included instruments of impact evaluation in which consultation with Indigenous peoples became mandatory if their territories are deemed a zone of impact according to the parameters of a project. What counts as impact and how to define impact returns us to the problems of groundwater governance and material politics. The improved and expanded Chilean state regulation, especially since these have come into effect in the north (since about 2015), has led to a further intensification of relations between industry and Indigenous communities. Thus, while regulatory and legal protections may have improved in character, the combination of geophysical and technical factors that allowed corporate groundwater exploitation to continue unchecked for many years have caused ecological exhaustion in some places. Moreover, engagements over impact between the Salar's Indigenous and industrial actors occur in ways that are inequitable, burdensome and morally fraught. It is the relations that are assumed in such engagements and which arise over time, the ways in which they have intensified with greater regulation, and the question of how to prevent ecological and social exhaustion that form the central concern of this book on the shifting and uncertain harms of extractivism and its advances.

A central problem in the conditions of 'advanced extractivism' is the incorporation of the discourse of both resource sustainability and valorization of Indigenous rights, while enabling the capitalist propensity to over-exploit. The dynamic between the extractive industry and Indigenous communities that is premised on corporate self-reporting and mitigation of environmental harm and the discursive recognition of Indigenous rights (especially to territory and self-determination), belongs to a broader analysis of corporate activity in advanced or 'late' capitalism and industrialism. Kim Fortun (2012, 2014) on industrialism and Elizabeth Povinelli (2011, 2016) on liberalism use the modifier 'late' in a similar way to Gidden's (1991) 'late modern' to refer to the social and structural dynamics of (apparently) unending capitalist political economic and social systems in which fracture and contradiction seems endemic. This is an apparent systemic failure of modernity, capitalism, industrialism for society and yet 'advancement' in the sense of planetary spread and inclusion in many dimensions of social and ecological life. This book demonstrates a number of dimensions of the ecological and social problems caused by relying on corporate self-reporting and voluntary environmental harm mitigation.

Three decades of 'engagement' between mining companies and Indigenous communities – under conditions of what I call advanced extractivism – reveal how extractive industries have engendered an increasingly complex set of relations, an intra-dynamics, of ecological, social and technical actors. A relation generally conceived in the past as characterized only by conflict (mining-Indigenous community) is cast as potentially smooth and coherent. By 'advancing', incorporating moral discourses about social and environmental sustainability, forging ever forward, and forming relations with social and environmentalist critics (Sawyer 2004; Kirsch 2014), extractivism demonstrates a significant capacity to uphold internal contradictions such as undermining and fragmenting opposition to extractive industry activity at the same time as trumpeting corporate ethics.

This late stage of, or advanced, extractivism is significant in a material sense for the southern end of the Salar de Atacama, which carries accumulating ecological burdens of industry, the social and economic pressures of consultation regarding those burdens, and, more recently, the moral pressure to enable lithium mining under the rhetoric of a necessary 'green energy' transition.

The use, abstraction and pollution of groundwater by mining is an engine for many social and ecological disputes on Indigenous lands across the planet. Advanced extractivism describes the intimacies and conflicts of transnational capital, the state and local and Indigenous peoples and ecologies, where their relations are significant to the process of extraction. The Chilean state has been a distant enabler of advanced extractivism, since the function and value of the extractive industries is embedded in social and political structure and process. Thus, the technical dimensions of groundwater as fuel for extractivism demand attention to the violent and slower sides of the mining industry. As for many Indigenous territories, the lives lived in the Salar are valued in capitalist terms according to the division between hyperarid surface and exploitable subsurface, but no matter how entangled in the industry, the division and other forms of capitalist schema do not exhaust the forms of being and aspirations of Atacameño-Likanantay people whose ideas, worldly analysis and social practice inform this book.

An Ethnography of Slow Resistance

Luisa told me, 'Sally, don't write about us as if we are anti-mining'. I sat across from Luisa at her desk in the offices of the Atacameño Community of Peine (in 2012). I had been telling her about an article I was drafting that emphasized a community critical analysis of Minera Escondida's environmental effects and social (benefits) failures. Luisa's comment indicates something of my failure to explain adequately to her how I was thinking and writing about local analyses of events and relations with mining companies.

Luisa was then in her early 30s, her father was one of the first in the village to gain work on the Salar for the lithium industry, and in the last decade, they have both been involved and sometimes outspoken in the community and for the territory about the fact of expanding mining industry activity, as well as the potential for mining-related opportunities (employment and other kinds of labour). Five years after that conversation, and at the end of a different period of fieldwork, I joined Luisa and others on a day trip from Peine to Antofagasta, boarding a chartered bus at 7 a.m. with more than 40 community members and children. When we arrived, we stopped by the beach and had lunch, and then assembled with protest banners and marched chanting through the city to join other Atacameño people from the region and those gathered by an environmentalist NGO outside the offices of the Regional Government. Inside the building, elected officials and officers from the Environment Evaluation Services (SEA) were discussing a new lithium mine proposed for the Salar de Atacama ('Peine NX Uno', then owned by Chilean Pablo Errazurriz). When we returned to Peine around 11 p.m. that evening, Luisa thanked me for joining their protest, saying 'it's a great support'. As Luisa made clear some years before, the Peine Community has never been 'anti-mining', but on occasion they have found good reasons and the groundswell of local opinion necessary to oppose new mining projects or raise objections against old ones. Their refusal to conform to either a pro or anti-mining stance, and insistence on a defining position as one that is in the interests of the community, makes none of their particular responses at any one time essential.

The Peine community members' diverse responses to the mining industry through time, as well as in terms of different community members' opinions, demonstrate an inherent, complex and unavoidable relation to their territories, including any industry actors installed in them. Their actions trouble the logics of advanced extractivism because they refuse convenient political positioning, neither conforming to a paradigmatic Indigenous resistance to mining nor entirely acquiescing to its terms. I use the term 'slow resistance' to refer to the fact of this troubled relation and the broader politics by which people seek to remain themselves in their responses to the complexities they face under conditions of advanced extractivism. The term does not seek to simplify the many responses to and analyses of mining by the members of the community of Peine in any one way, or encapsulate as 'resistance' every community member or leader's approach to relations with industry.

The term slow resistance is a counterpoint to what Rob Nixon has termed 'slow violence', the environmental and social harms from industries such as mining that 'occur gradually and out of sight' (2011: 2), with temporal and spatial dynamics that are incremental, accumulative and wrought by the structural inequalities of global industrial capitalism. The effects of slow violence not only fall disproportionately on emplaced people (Indigenous Peoples) and 'the poor,' but as I have argued elsewhere (Babidge 2019), the unspectacular effects of slow violence are managed by the industry through company science, a construction of ignorance that aims to disable public knowledge of and organized response to extraction-associated environmental harm. In Nixon's book title, he identifies the counterpart of 'slow violence' as the 'environmentalism of the poor'. However, few Peine community members I listened to would cast themselves as poor, nor as environmentalists.[11] Under the conditions of slow violence, the incremental refusal by community members of company environmental knowledge claims, in corporate reporting and environmental impact assessment studies led by consultants in service to the industry, constitutes an unspectacular refusal to see themselves as wholly subject to these conditions. This refusal comes in a range of forms, and is explored throughout the book. I briefly outline some aspects of what I mean.

The village of Peine, with a permanent population of around 400 Indigenous Community members, plus a small number of non-Indigenous permanent residents, and a floating populace of mine workers and contractors that at times may double this number, is historically entangled with the extractive industries through the economic and presential dominance of the industry in the Salar, in their village and across the Antofagasta region. Throughout the twentieth century, many Atacameño men and women, and sometimes whole families (or some members thereof) took up a permanent residence in the regional city of Antofagasta or Calama, where waged work, especially in the mining industry, provided a cash income and educational opportunities for their children. Others remained in Peine all of their lives, also drawing on the resource of a home in the city when necessary.

Mobility between the village and the city has long been simply how people live. While a person may reside mostly in Calama, for example, they travel to Peine for family events, school holidays, work in the planting or the harvest or to participate in religious or other annual festivals. Peineños living in the village travel by bus or private vehicle to attend shift work or college, deal with state authorities or banks, purchase household goods, construction or agricultural materials and tools, costumes or other goods for festivals. A small handful of Peineño men live in mining company dormitory camps, such as the one owned by SQM (located on the fringes of the Salar fifteen kilometres north of the village) during the days or weeks of their shift. The mining infrastructure spatially surrounding the village and mining associated work is internal to bodies and imaginaries of future possibilities. It is also on the surface of material life, such as the robust cotton drill workwear worn in the mining plant, and outdated kit ubiquitous in agricultural labour. The community is bound with the history and other temporalities and materialities of advanced extractivism, but this does not exhaust their social reality. The same is true for their ecological relations, which at times appear to complement a conception of nature-as-resource embedded in extractive capitalism or an interest in market economics, but also sit outside such configurations in the everyday manner in which members of the Peine community relate towards one another and others.

The circumstances of mining and consultation about the effects of mining in Indigenous territories sees mining company employees charged with developing social relations with Indigenous Peoples, who are in the position of responding to the imposition of extractive capital and reflecting on what counts as socioeconomic benefit and ecological impact. The values that inform 'what counts' as impact are crucial to the operation of advanced extractivism in Indigenous territories since this objectified confrontation of worlds implies for community members a moment of enforced reflection on 'the self' as community. Due to the fact that consultation happens with 'the community' (and, as I outline in the following chapter, this is the norm and expectation), members of the community must therefore consider developments *vis a vis* a collective self ('the community') and their own desires (see Beckett 2013). This reflection on the multiple 'self' in action is intensified (but not unique) to the advanced extractivist context since it involves sets of questions on the future moral good (as 'development') in collective terms, and in terms that 'should' reflect an authentic response *as Indigenous peoples*. An explicit (and shifting) position regarding 'the community', and its values and desires in relation to 'the company' is only generated through significant labour. Chapter 7 considers the intensity of this labour and its social burden.

The boundedness of community/company is neither empirically nor conceptually well-founded (e.g. Golub 2014; Welker 2014). However, it is from such a dialogic (or 'entified') understanding, as well as the dialogue of ethnographic encounter that I draw the term 'slow resistance'. As Tedlock (1991) argued, anthropology has shifted from participant observation to the observation of participation, in which the ethnographer and her interlocutors together are the subjects of the narrative. In this vein, the representations of people's worlds are intersubjective in the sense of Duranti (2010); not an achievement of understanding, but evidence of an effort to put the self in the place of the other. In the chapters that follow,

I seek to describe the ways I observed actors from 'the company' and 'the community' relating and communicating as persons and also as contesting entities.

As a phrase, slow resistance runs the risk of contradicting Luisa's admonishment, as well as problematically focusing on Indigenous alterity (Trouillot 2001). Certainly, the term resistance calls up a utopian other (the savage slot, the moral native, etc.) and risks anthropological disappointment, a hope dashed (Greenberg and Muir 2022) if one is to seek a future in which resistance prevails. However, Trouillot's optimistic critique of anthropology in globalizing conditions was that in accounting for the anthropologist's (the West's) concepts and categories and how these are founded, we might produce works for the complex and globalizing present that retain the most useful of the discipline's foundations. As such, my use of the phrase 'slow resistance' seeks to narrate the dynamics of Atacameño-Likanantay people's practice in relation to their (extractive) others and related selves, as well as write about this as anthropological witness to the powerful work of transnational capitalism.

The globalizing idea of corporate social responsibility, part of the slow violence of extractive capitalism (Gamu and Dauvergne 2018), was introduced in structured ways in this location, where ecological effects of industrial activity have been muted by the geophysical landscape, technical monopolies on data about impact, and regulatory partiality for mining. Thus, the pairing of slow violence and slow resistance neither elevates 'resistance' to the utopian, nor seeks to forget what people in Peine taught me to hear; the practicalities and historical realities of extractivism in their territory demand from them varied responses, generated not only in 'external' factors, but always by their own realities.

Indigenous social and ecological practice that articulates life in relation to other beings and territory in terms of the potential for a future life is in contradistinction to the value of that same territory as resource for extractive capital. However, the first is not always opposed to values that may be derived from the second. This is in no way unique in the planetary story of the mining industry and its challenges for Indigenous and rural peoples. What is (I find) compelling about the southern Salar are the modes of people's social response in regard to the powerfully muted history of extractive relations and the always emerging logics of what I characterize here as slow resistance. My perspective, it must be said, differs from regional commentary, and others' external analysis. To further my argument, I draw on ethnographic material that expands upon slow resistance, a multidirectional set of relations and histories of practice in parallel to, as well as in the refusal of the logics of extractivism. As such, it both incorporates and exceeds the idea of development as a teleology of progress and relies on technologies of advanced extractivism and its distinctions between surface and subsurface, as well as demonstrating a story of incremental and sometimes problematic economic flourishing fostered by extractivist activity.

Diversity within any community of peoples, what Borofsky (2020) calls 'intracultural diversity', and the question he points us toward – how is diversity organized? – orients us to Luisa's warning about representating her community. Considering the organization and negotiation of sociocultural forms must begin with the always-extant fact of difference. In Chapter 1, I discuss in more detail my research methodology and the ways in which people in Peine manage 'community' analysis and opinion in relation to outsiders. In conversations I elicited or the dialogues among community members in which I participated, I sought to assemble with Atacameño people and others a sense of the diversity of their responses to the questions of advanced extractivism. As I was asked to do by community leaders, I discussed in community meetings, individually, in conversation with those like Luisa, or among groups of people over a meal or other social gathering, the material I was writing, and in doing so invited analytical conversation. In each of these different settings, it is members of the community who would challenge me to explain how I saw things and what my conclusions might be. My slow ethnographic attempts to respond well to people's questions about how I was representing things in my writing about the Salar de Atacama rely on many different opinions among community members in Peine. People corrected me when they had decided I had misunderstood, and invited me to attend, listen to or record particular events in order to understand, which prevented me from accessing certain matters, provided different angles of argument or opinion and directed me to speak with those who might inform me further, especially where this would challenge or deepen my understanding of relationships between the community and the mining corporations in their territory. This ethnographic research is built on that incremental dialogue and is unfinished in the sense of an ongoing conversation.

The Flow of the Chapters

An interest in social change is at the core of this ethnographic study. It is not a chronological telling of the stories I heard from people about their pasts, or the histories I sought out in documents. The chapters tack backward and forward in time, following how things that matter were told to me or became relevant to an understanding of the present. Nonetheless, the book builds toward a final chapter in which the current 'overburden' of mining projects threatens to exhaust the potential for a future good life in the Salar.

In Chapter 1, I expand on the analytic of 'material politics', giving this texture in terms of everyday economic and ecological relations. The ethics of relations refers to a parallel and apparently contradictory sense of ecology in the capitalist idea of ecosystems management and the ecologies of relations in Atacameño territory. Both are vibrant, contested and are articulated alongside one another, causing trouble. The chapter begins with a story of infrastructure (the stadium) that instantiates relations between mining companies and Indigenous communities. It then sets out three ethnographic sketches of material relations in Peine, the first in regard to processes of community decision-making in *asamblea*. This, too, sets out the terms by which ethnographic relations have been possible. The second is an account of the ecology of value generated through water labour, which also details some of my own labour, and which gave rise to key aspects of an analysis of social life in Peine. Finally, a historical account of 'development' as it arose in the stories Peineños told me during oral history interviews is contrasted with development as it is envisaged under conditions of advanced extractivism.

The place called Tilopozo, mentioned in the opening sentences above, is more properly introduced in Chapter 2. It is a rare location featuring surface freshwater in the southern Salar. As such, it is an environmental indicator of the effects of the extensive groundwater extraction by copper and lithium miners. Tilopozo is lush with community history and a conspicuous subject in plans for making a good life for future generations. What I refer to as the ecological histories of relations between water and people are a reference to water futures that are imagined despite the threat of extractivist exhaustion. The chapter is structured by the multiple materialities of the groundwater that nourishes Tilopozo; as enigmatic geophysical material in aquifers, as sets of numbers in environment reports and registers, as state-defined quantities of a commodity and as essential substance for life in place. Focusing this chapter on Tilopozo as a subject is an argument for the revelatory potential of apparently insignificant and 'remote' ecologies, an exploration of how groundwater's enigma generates a certain politics and how such politics may be shifted and refused.

The ethnographic examination of the intersection of ecological relations and the expansion of vertical technologies of advanced extractivism demand a brief return to one of the beginnings of relations between transnational mining and the Peine community. Chapter 3 is centred on an epic story set mostly in the 1990s, about the struggle to gain a new source of domestic freshwater for the village. Using oral testimony and local archival collections, the chapter emphasizes the multivocality of domestic water's history, as it was told whenever I asked about social change associated with the mining industry. For local narrators, it is sometimes a cautionary tale; communal labour and self-reliance harnessed a spring for domestic water when the state failed to provide, and seeking financial help from mining companies marked the beginning of intentional relations with global extractive industry actors. As this chapter shows, many kinds of material relations are at work in the provision of freshwater. In building the road, digging channels and laying pipes, labour materially affirms villagers' age-old solidarities and visions of development on local terms. As a history of relations with others, including mountain entities, neighbouring Indigenous villages, the mining industry and the state, it articulates the ecological and social relations that power community authority and slow resistance.

In Chapter 4, the agreement as a technique of globalizing 'ethical capitalism' is put to the test by the fact that the Peine community demands that mining companies negotiate agreements for development benefits with them, and the ways in which the community has refused to accept agreements with extractive capital as instruments that reduce their political agency. The chapter examines how people described their experience of the relation generated by the community agreement with Escondida and shows how this is focused on their assessment of different distributive successes and failures of 'development benefits'. The details of different actions related to and imaginaries of 'development' that are associated with signing agreements provide the contours of such operations, as well as the beginnings of advanced extractivism. For Escondida and Peine, one result of these agreements was the heightened awareness of the disjunctive moral economy of the relations of contractual agreement, nourished by the unknowns of groundwater and suspected effects of company extraction.

The labour and life histories elicited from older Peineños who long worked with the salt ecologies of the Salar and in the early lithium industry narrate their efforts to make the most of these sources of a good life within their territory. A character sketch of lithium plant boss, Don Tomás, traces the first lithium company's patronage of the village of Peine, and with it local hopes for prosperity. The beginning of lithium mining near the village in the early 1980s was a significant period of ecological, geopolitical, historical and socioeconomic change, just as the contemporary lithium boom is generating uncertainty and flux. Driven by transnational capital, both periods are also inescapably enabled by the state. When I began research for this book, people in Peine did not speak about lithium mining as threatening to produce irrevocable lifeworld loss, but the expanding lithium industry on their doorstep has changed perspectives on the past, as well as perspectives on the material politics of this industry for the future.

Under advanced extractivism, companies manage the political and ethical complexity of local tensions arising from mining impacts of various kinds, in the intentional process I describe in Chapter 6 as 'making relations'. Relying on return and in-depth interviews with mining company employees from copper and lithium companies, as well as insights drawn from mining industry 'social responsibility' conferences, this chapter addresses the question concerning how extractive industry *actors* enacted the 'responsible business' of mining in the Salar de Atacama, and what its effects are. The problem of company social relations employees in their desire to 'do good' and mitigate harms is that they are unable to both enact company policy and be consistent moral persons in their relations with community members. As an advanced extractivist technology, 'making relations' entails its own violence. The chapter explores the interpersonal techniques by which extractive capital seeks to secure a conflict-free extractive zone, is an analysis of the social, ecological and moral failings of advanced extractivism for those employed within it, and an account of Indigenous community rejection of *some* companys' agreement making in the Salar de Atacama.

Since 2016, an upsurge in Atacameño-Likanantay community-led protests and administrative refusals against copper and lithium extractive industries in the region has occurred. These are in response to the expansions of the lithium industry, attempted extension of groundwater pumping by Escondida and Zaldivar, threats of new mines, and in parallel, increased material recognition of Indigenous rights and increased state action on environmental laws. The mining boom for lithium has seen ecological harm to the Salar de Atacama become tangible at the surface of the landscape and in all the textures of daily life. Chapter 7's backdrop is the rising public conflict over extractivist harm driven by the boom in transition minerals and an increasing role played by the state as a fragmented regulator of extractivist activity. In the past five years or so, Peine's strategies to respond to extractive pressures, using regulatory and legal mechanisms at their disposal, have led to the cessation of new mining projects. Their tactics have also seen communities gain significant funding from industry for local life projects. However, the burden of revealing the actual and potential ecological harm of extractive industry activity falls most heavily on Indigenous communities. This 'overburden' entails complex social effects and threatens the fragile and relative autonomy that people rely on in determining their futures.

Notes

1. *Aflorar*, to surface or to emerge, especially a mineral body or flow of water (Real Academia Española, https://dle.rae.es/aflorar?m=form, last accessed 19 June 2023). 'Geraldo', as with all first names that appear in the book, is a pseudonym.
2. The characterization of water's 'intelligence' is rhetorically reminiscent of the ideas of Chilean biologists Humberto Maturana and Francisco Varela (1980), whose theory of autopoiesis in part considers that all living beings and systems are cognitive self-generating systems (my thanks to Sol Rojas Lizana for pointing me to this parallel). As philosophers of science Prigogine and Stengers (1984) have argued, the sciences may be traced as moving toward a conception of the real that is less about being and certainty than becoming. Company scientists reduce the animation of the world (water) to a uni-directional instrument, rather than the contingent becoming apparent in Maturana and Varela's conception, or the 'chaos' of becoming in Prigogine and Stengers'.
3. Indigenous recognition legislation (the Indigenous Act 19.253, 1993) made this process possible. Anthropologist Marinka Núñez documents much of the work undertaken (2011, 2015). Lawyer and anthropologist Alonso Barros (e.g. 1997, 2006), also provides a detailed review. Archaeologist Fernanda Kalazich has documented in maps, oral history recordings, and participatory research Peine's territorial and cultural heritage (e.g. Kalazich 2015). Raúl Molina (2017) documented Atacameño people's travel and trade throughout this sector of the Andean cordillera through time.
4. Chilean historian, Ricardo Latcham's early critical analysis of the north American intervention and enclave system (1926), Finn's (1998) study of copper culture, recent work by anthropologists Anita Carrasco (2020) on Atacameño entanglements with the mine and its water use and Marina Weinberg (2021) on the bodily-mine generation of wealth are only some examples.
5. The myth of Chilean exceptionalism has been considered, for example, in terms of racialist ideologies (Walsh 2019), national imaginaries regarding geopolitics of remoteness (Salazar 2013) and political economic history (Stern 2006).
6. In a report published at the end of the civic-military regime administration, copper production under CODELCO was reported to have grown by 95 percent between 1973 and 1989. CODELCO's involvement with the manufacture and selling of copper in partnership with the Europeans and Chinese was reported as not only having enabled significant economic growth, but also technological development: 'If we add that CODELCO-Chile is exporting its own technology to different parts of the globe, including the United States of America, then our investment is making an entrance into a scene to which traditionally only developed countries have had access' (Gobierno [Militar] de Chile, c.1990: 238). This vision of economic and technological development is nourished by extractivism and is part of the imaginary of copper nationalism in Chile (e.g. Moran 1974).
7. In an article published on the cusp of the Estadillo Social, authors Ruiz and Caviedes (2020) refer to Chile's 'advanced neoliberalism', and detail the extent of capitalist functioning in the public service, privatization of everyday life (water, pensions, education, health and welfare) and the extreme forms of concentration of capital in the hands of the few and resulting forms of inequality.
8. Document available online: https://www.senado.cl/logran-historico-acuerdo-para-nueva-constitucion-participacion/senado/2019-11-14/134609.html (last accessed 1 April 2024).
9. The Water Code Amendment Bill seeks to make significant doctrinal changes, but it is too early to say how these will affect ongoing water inequalities.

10. Whereas Gudynas (2012) excludes Chile from neoextractivism (it is a 'different case'), Svampa's (2019) analysis clearly draws it into the schematic.
11. While the socioeconomic status of families and people in Peine may indeed fall within a measure of poverty, people told me that unlike those living in cities, they owned their homes, could grow food and feed their herds and enjoy (parts of) their territories and waters. All were subjective measures of the good life in the present and its future potential.

Chapter 1

ECOLOGIES OF ADVANCED EXTRACTIVISM

The morning sun was warm on our backs and the cans of beer in our hands still cool. I joined others in a hastily arranged *convido*, a ceremony in which the presiding *cantales* (ritual leaders) invited the volcano peak Llullaillaco, the Pachamama and ancestors to attend the opening of the Llullaillaco Stadium in Peine. The sports and event venue was named after the peak that Atacameño-Likanantay peoples of the Salar de Atacama refer to as the chief of water. Llullaillaco is located at the southern point of the Peine community's claimed indigenous lands, and presides over the Salar de Punta Negra, from which copper miner Escondida has pumped groundwater for some decades. In 1997, Peine and Escondida signed an agreement for 'development benefits', and the indoor sports stadium was one of the first major projects funded. Twenty years later, the Llullaillaco Stadium was to be re-inaugurated with a lavish day-long festival, and the community president, some community members and a handful of company employees had gathered at the rear of the newly re-finished building for a pre-event *convido*. The *cantales* asked the peaks to bless the building and the *pueblo*, and to ensure that the waters these peaks oversee keep flowing and the skies keep making rain.

At the front of the stadium, other members of the Atacameño Community of Peine, employees from Escondida copper mine, its global operator BHP, and its social projects trustee, Fundación Minera Escondida (FME), were assembling under a large marquee for the day's festivities. By the time the small *convido* was completed, a crowd had formed and formal proceedings began. The national anthem played, then the President of the Atacameño Community of Peine spoke expansively, naming each of his ex-President predecessors in turn, paying tribute to those whose dedication had initiated the stadium project two decades before, and all others who had endured disagreements with the mining company over aspects

Notes for this chapter begin on page 55.

Figure 1.1. View of the village of Peine looking west, with the Lllullaillaco Stadium apparent. © Sally Babidge.

of construction and part completion in the twenty years since, during which the huge structure was used sporadically. He finished by thanking everyone local, *los vecinos*, then the Mayor of the Municipality of San Pedro de Atacama, and finally and simply, the company ('*la empresa*'). Their speeches then followed in this order.

As a foreign mining company, BHP has enjoyed a significant public profile in Chile partly because of the work the parent company and its philanthropic Foundation (FME) has done to publicize its record of social and community investment in the national economy and mined areas. The company cites itself in industry practice guides as a 'good example' of mining CSR programs in Chile and elsewhere. It has built stadiums throughout urban and rural localities in the Antofagasta region like the one in Peine, as well as financed social infrastructure, welfare programs and partnerships with government for public building projects outside of the region. For many years, it has also been the subject of public criticism relating to environmental harm from groundwater extraction.

Reflecting its name, Escondida (lit. 'hidden'), the mega-mine operated by Anglo-Australian BHP (majority owner in a multinational conglomerate that includes Rio Tinto and Mitsubishi) is located around 130 kilometres to the southwest of the village of Peine, at 3,500 metres above sea level, one of many open pits dug into the coastal range.[1] Between 1986 and 2001, Minera Escondida Limitada or its operating company purchased rights to extract groundwater in volumes totalling more than 3,000 litres per second from two principal areas in the Antofagasta region; an aquifer partly located under Monturaqui, to the south of the Salar de Atacama,

Figure 1.2. *Convido* for the re-inauguration of the Lllullaillaco Stadium, 2017. © Sally Babidge.

and from around the Salar de Punta Negra, at the foot of the Llullaillaco peak.[2] Escondida/BHP uses such mega-volumes of groundwater to process more copper than any other mine of this kind in the world.[3] A 2018 'Water Report' entailing a company audit of the company's water use across all operations (Australia, US and Chile) reports that the Escondida mine used greater volumes of water (both groundwater and seawater) than all of BHP's other seven operations combined, totalling around 170,000 megalitres (BHP 2018: 36). The mine's extractive infrastructure includes coastal desalination plants in Port Coloso south of Antofagasta, designed to supplement and replace the many thousands of cubic metres of freshwater that Escondida's operations demand annually for washing ore and machinery. The plant is offset infrastructure: its material form and function are a response to Chilean state and public demands that the company address the threat to ecological life generated by the company's groundwater pumping in the Salar de Punta Negra and the southern end of the Salar de Atacama. Escondida's expansive infrastructure is not the only extractive activity in the Salar. Copper and lithium operations jostle for ongoing and new access to the groundwater associated with the Salar de Atacama (and every other groundwater source in the area), with many thousands of litres per second drawn from aquifers below the arid surface. Nonetheless, Escondida is a giant among global extractive giants, operating on thirsty technology in a hyper arid, remote and high-altitude desert region.

At the stadium opening festivities, the Director of FME consented to a short interview. I asked him, 'what would an event like this achieve for the company at this juncture?', to which he responded in part,

> This is a beautiful settlement, for a work that symbolizes the *trabajo en conjunto* [joint effort] between the company and the community. What we want to do is *trabajar juntos* [work together]. Each one has their objective, but in the end, we can't be on different sides, we have to converge. There is a phrase that Indigenous communities use to refer to *'el buen vivir'* ['the good life/living well']. *El aprovechar de la naturaleza* [Making the most of nature], um, that we are *parte de ella* [part of it] and we are forging ahead, to construct a future together.[4]

Given both the specifics of Escondida's extractive operations that I outline above and a general knowledge of copper mining's terrible environmental record, I would not be alone in finding 'making the most of nature' a curious translation of 'an Indigenous concept of the good life', or that 'together' describes well the mining company-community relationship. Latin America is famously envisioned by Eduardo Galeano (1971) in terms of a body bled of life over 'five centuries of pillage' by the colonial powers of extractive capitalism. The mining industry is renowned for the effects of the 'resource curse' for those living local to industrial operations (Bebbington, Hinojosa et al. 2008; Langton and Morel 2008; Weszkalnys 2011; Acosta 2013; Gilberthorpe and Rajak 2016). Empirically grounded ecological and social critiques of mining generally, and copper mining in particular in the Antofagasta region, are almost as rich as the mineral and hydrological veins plundered. At the same time, those with experience of contemporary extractive industries may not look twice at a company water-use report that refers to its mega-operations of groundwater pumping in terms of water 'stewardship'. Indeed, while I hoped to gain some insights into his views on the ongoing relationship between company and the community, I had anticipated eliciting in conversation with the FME Director at least something of what I have heard referred to in Peine as 'company propaganda' and what Kirsch (2010) refers to as the language of corporate oxymorons such as 'sustainable mining'. Passing off the FME Director's utterance about 'being one with nature' as obscurantist dissembling does not prevent the efficacy of such discourse in the reproduction of the harmful conditions of extractivism.

Extractivism, as a term of critique, refers to a political economic model in which earth, water and minerals are, first and foremost, *resources* which may be transformed into commodities, and which thereby generate the possibility of 'development': a moral economic good of growth and, by extension, social modernity. 'The gift' of modernity is a powerful metaphor in extractive development (Rajak 2011a). With the diminished functional operation of the state in Chile's neoliberal capitalism, as well as elsewhere (neighbours Peru an important parallel, see Silva 2008), transnational corporations such as BHP pride themselves on conferring development in communities and regional cities as part of their policy of corporate social responsibility (CSR). Since the late 1990s, transnational mining capital has drawn on a moral discourse of development to 'see like a state' (Scott 1998), formulating policies of CSR and 'environmental sustainability' to generate the appearance of support to Indigenous community and order in the management of the environmental 'impact' of company activity (Karliner 1997; Jenkins 2004; Yakovleva 2005; Smith and Helfgott 2010; Kirsch 2010). Company employees like the FME Director, responsible for CSR implementation through an external philanthropic foundation, express the fantasy of an ethical and ecological capitalism (a *better* development) in his desire for a 'convergence' of the different parties' values for development. This fantasy integrates a knowledge of liberal values such as the recognition of the rights of peoples and populations, and the value of environmental governance. Public experience and knowledge of the industry's actual effects is so widespread one might expect the speaker himself to recognize his own hyperbole.

Enacting responsibility, pronouncing on company respect for Indigenous rights, and producing company data to assert the management of 'resources' is part of extractive business, despite and because of company harm. Local and global projects that articulate dissatisfaction with the mine or other extractive activity can be interrupted, slowed or re-routed by a range of corporate measures of community benefit delivery. Company employees contribute at the local level by working on community projects such as infrastructure (stadium) planning and including community members in company environmental monitoring activities (cf. Velasquez 2012; Welker 2014). Company engagement with Indigenous peoples helps enable extractive capital to continue extracting since evidence of this activity, such as photographs of community events or tabulated data of funding provided to communities, can be used for shareholder brochures, providing the appearance of good corporate conduct. While such endeavours have been labelled 'green washing' and local peoples are not duped by the rhetoric and are, at times, affronted by being used as fodder in CSR campaigns, people also seek the infrastructure or other material benefits that companies promise to deliver.

Critiques of development (Ferguson 1994; Escobar 1995) are well worn (see Gardner and Lewis 2015, 61–65). However, rejecting the efforts at or actual projects of development appears as a rejection of the idea that people deserve better material conditions for a good life. As Korf (2023) argues, development's 'moral excess', the fact that one cannot *not* want it, is the impossible characteristic of development that makes many continue to work for and desire something despite knowing very well its failures.

The FME Director's claims to work together with the community and my description of the events of the Llullaillaco Stadium opening narrate some of the shifting material and moral relations of contemporary extractive development. These are the negotiations between company and community that are the focus of all manner of events discussed in this book. Company financed 'development' materialized in social and environmental projects shows how the failings of advanced extractivism parallel the failings of 'development' writ large, but also the how extractivism continues to advance amidst widespread critique about harm. As managers of company relations assert the material bases for life (economy and ecology), there is in parallel lively scepticism on all sides. How then do such claims of 'good capitalism' get sustained? What people in Peine do with globalizing discourses of economic and ecological relations, and how they have dealt with other projects of modernity in the past, matter for thinking about the power of extractivism and the cracks in its apparent mastery.

This chapter narrates three ethnographic descriptions of Peineño material and moral life and finishes by returning to the questions of advanced extractivism. First, the *asamblea* as normative and practical locus for 'community' decisions, provides the grounds for the terms *asamblea* and community as they are used in this book. Given the authority of the asamblea, this account is also about my research relationship with the community. Second, I narrate my experience of water labour associated with agricultural land, irrigation and ritual practice, and what participating in these activities made evident to me about water relations and ethics. Third, I provide an account of the various stories that illustrate the idea that people in Peine have always been interested in 'development' and modernity, and the character and materiality of these ideas. Finally, I return to the vignette of the stadium opening and to questions of development and ecology in late capitalism.

'La Asamblea Manda': Community Decision-Making, Politics and Morality

The phrase *'la asamblea manda'*, literally, 'the assembly decides', refers to a moral imperative: individuals and elected leaders of the community *should* not be the ultimate decision-makers; instead, all decisions that affect members of the community must be taken by the *asamblea*. The *asamblea* is an event (an assembly of community members) and community polity, the decision-making body in regard to matters that pertain to the Indigenous community. It is the location for legitimate political and social action over territory and water, including decisions by the Peine community about their territories and waters, and between them and other Atacameño communities and organizations, mining companies, the state, and other non-community members seeking relations with the community, such as researchers. The *asamblea* as a body elects the community president and *directiva*, or leadership committee. The *asamblea* organizes and is an instantiation of the achievement of 'community' given Peine's diversity of people and their different opinions. Wolf's (1957: 2) comparative concept of the 'closed corporate community' as a group of livelihood generating agricultural producers ('peasants') who 'maintain a body of rights to possessions such as land . . . [and] prevent outsiders from becoming members of the community, [and maintain] a perpetuity of rights and membership' is a reasonable (if a little dated) description. I refer to it in consideration of Wolf's later (1986) discussion of the concept, in which he emphasized that he had sought to reflect 'local and parochial relationships in terms of wider unfolding economic and political processes', while 'simultaneously to grasp how [people] responded to these [wider] processes through culturally informed action and action-involved cultural forms' (328). The idea, then, of the term 'community', and my emphasis throughout the book on community as a social and political achievement, aims to consider the entity not as an artefact but as a set of relations both changing and objectified (cf. Smith 1989). As such the idiom of 'community' points to a structural fact, useful for insiders and outsiders.

The *asamblea*, as an event, refers to any meeting called by the community president and *directiva*, held in the *sede* (lit. 'headquarters', meeting hall) to which all *socios* (registered adult members) of the Atacameño Community of Peine (incorporated under CONADI) are expected to attend. To underline the expectation, those in attendance sign a roll. In a special *asamblea*, roughly annually, *socios* nominate candidates for new leadership, and each *socio* then casts a single a vote for their preferred candidate. The president is she or he who receives the most votes, with treasurer, secretary, and two ordinary directors according to descending vote counts.[5] The *directiva* emerges from the *asamblea* as the changing articulation of Peine's polity and of Peine's politics.

The *asamblea* as a site of community expression, internal negotiations and reception of outsiders means that it has been a principal site of this ethnographic research. I have observed and participated in over thirty between 2011–2019.[6] The following discussion serves to explain the ways in which the *asamblea* as polity directed my ethnography and therefore gave rise to the particular truths that I commit to these pages.

After preliminary conversations with individual *socios* in 2009, I planned a six-week period of fieldwork (November–December 2010), during which I would seek permission for a three-year ethnographic project. I rented a room from Mateo, whom I met the previous year, and hand-delivered a letter to the president and *directiva* in which I petitioned them to raise my research proposal in the 'next *asamblea*'. I was warned by the former president that it was a difficult time of year, with many other events going on, but he assured me that it was possible that a meeting might happen during the month of December. It did not eventuate. Nonetheless, the *directiva* alone signed a letter of permission allowing me to stay in Peine and spend time talking with community members during that initial six-week period, and on the condition that any research data I generated would be destroyed if the full *asamblea* did not later agree to the study.

Early the following year, my letter was read to the *asamblea* (in my absence) and my request was authorized. I returned in July 2011 and each subsequent year (sometimes twice annually) until 2019, to follow up on unanswered fieldwork questions, discuss new research directions, analysis and findings with each newly elected *directiva* and (if I was asked to do so) sometimes the *asamblea*. From 2012 onwards, members of Peine's *directiva* began to request I undertake small jobs for them, and from 2016 these jobs were more formal requests to analyse mining impact assessments, or providing access to ethnographic data (mine and others) in support of community responses to mining activities.[7] Fieldwork that generated data for this book was authorized by each new Directiva at the beginning of periods of fieldwork that lasted between six weeks and three months, until 2019 and published articles were provided to the Directiva (some fully translated to Spanish).

In late 2022, I returned to Peine (for the first time after the COVID pandemic) to request an audience with the *asamblea* to discuss the draft of this book. Like the very first period of fieldwork, the community was preoccupied with end-of-year (Christmas, end of School year, etc.) events. And as in other communities of the Salar, Peine's population had expanded in the years of the pandemic; those who resided in the city for many years returned permanently, constructing new homes and reviving old garden plots with new crops. My friends and colleagues in the village told me, 'everyone knows you, Sally', but the mutual relations we share do not represent the reality of the changing polity. Our alliances are based on the years of field visits and remote phone calls, and when co-present, we have laboured, eaten and shared questions and thoughts about their worlds and mine. But these relations and friendships do not extend to all members of the community and the community president and *directiva* were dealing with mounting external demands, including from researchers such as me. I had never met the president, and he told me, 'we don't know if you will sell information to mining companies'. He chose to represent the interests of the community by refusing to enable my audience with the *asamblea*, and so no new ethnographic material was generated during that time.[8]

Over the decade of fieldwork in Peine, I have witnessed the president or *directiva* implementing a new project they envision as beneficial by taking decisions without having passed these by the *asamblea*. As a moral accusation, '*la asamblea manda*. They [the president or directiva] shouldn't take decisions alone',[9] articulates an ethic of the community as an achievement of political unity, not through the divestment of decision-making powers to elected members, but by investing in the responsibility of the elected members as those who will bring important decisions to the *asamblea*. Of course, what counts as an important decision is subject to the judgment of the *asamblea* and *directiva*. In the chapters that follow, I show how it was Peine's *asamblea*, not its leaders alone, who refused the terms of 'transparency' presented to them by the mining company Escondida and insisted that negotiating with mining companies and receiving 'development benefits' is not the same as agreeing to the terms of company-caused environmental harm (Chapter 4). As Peine's *directiva* and other leaders (those without an elected position but sought for their informed opinions) made arguments about mining activity and the pressures of consultation and participation as 'overburden', these were formally supported by the *asamblea* (Chapter 7). In Chapter 6, I show how mining company employees charged with 'making relations' have attempted to influence the *asamblea* through pressure on members of the *directiva*, and how members of the *asamblea* have moved to prevent such interference in local affairs. The *asamblea*, as a decision-making event and a set of changing community relations, can be a political force for expression and a site for hashing out the diversity of local opinion. It may also be a temporal brake on outsider requests for a community decision. The *asamblea* is the collective expression of a changing polity, contingent on people, time and place.

Water Labour and Ecologies of Slow Resistance

Membership of the Atacameño Community of Peine is possible if a person can show they have ties to ancestors associated with Peine's village lands and waters. In the articulation between community territory and the agricultural fields inherited by persons from family members is the play of the social dynamic between the individual and communal. The older residential parts of the village characterized by ancestral dwellings and familial lineages may be traced in spatial clusters of these homes (M. Núñez 2015: 137–39). Newer areas of the village expanding upward and south are not patterned in the same ways. Those wishing to construct a new home may request from the Junta de Vecinos the right to a specific plot of land, and thus also rights and support to connect to water and electricity grids.[10] It is forbidden to build homes in the *monte*, the cultivation sector that stretches along the northern side of the village, and impractical to build in the *playa* or the sandy lower western edge that borders on the Salar, and where the agricultural and flood waters run out. Conforming to Wolf's (1957) 'closed corporate community' (as discussed above), non-members of the Atacameño Community of Peine (the Indigenous community) cannot own land or waters, and with a handful of exceptions (all of whom are spouses of Peineño women) this is true.

Agricultural land and rights to the waters that irrigate them are inherited and inheritable property, not directly associated either in space or in title to a familial home and bequeathed while still living by individual parents to offspring of both genders, or from grandparents or parent's siblings. Given the many years of the division of land, each *propietario* or landowner may be responsible for *melgas* in multiple sectors of the *monte* (the agricultural lands).[11] My ethnography to some extent bears out the findings of others (e.g. Bennett and Zwarteveen 2005), who write that women and men's involvement in irrigation, water politics and water struggles are marked by gender, but not divided. *Melgas* are held by individuals and are the subject of personal responsibility and rights to irrigation water for the crops they choose to plant (dominated by maize and alfalfa). A *propietario* might decide *not* to plant the fields she inherits, and *melgas* may be left fallow for years (and if a *propietario* takes up a residence in the city to school her children, for example, potentially for a generation or more). Those with adjoining fields refer to one another as *vecinos* (neighbours), just as they might with others whose house adjoins theirs. *Vecinos* are in frequent communication with one another if they have planted their fields, especially during the irrigation months.

As fieldwork periods each year almost always fell between June and January, I took part in many days of planting with three different families of *propietarios*, and worked with six different women while they flood irrigated their fields. In November 2015, I spent time following the irrigation water each day, trotting alongside the canals as the flow rushed downhill and tagging along with Justo, who was *'juez de agua'* (lit. water judge). *'En eso estamos aqui regando'* (here we are irrigating), people responded in what I find to be a rather existential fashion, as I showed up with Justo and asked the landowner how things are with the water. 'This is what we do', irrigate.

The '*juez*' manages the system of the three reservoirs and main canals, opening the correct reservoir early, and following the water to its destination; the fields of the person directly upstream from their neighbour having been irrigated the day before. Mateo told me over lunch one day that he considered Justo '*muy exigente*' (very strict), opening the reservoir early, present at each area of *melgas* to take the water, and charging fines as appropriate. During the irrigation months (September to May), *vecinos* have time to speak with one another and the *juez*, and trade witticisms and complaints as each *propietario* reaches their garden in time to receive water into their *melgas*. I scooted along behind Doña Inéz as she hurried from her home and through *el monte* to arrive in time to take irrigation flows into one section of her fields. We passed by and greeted people working to '*aporcar*' (tilling the soil up around the stems as a support against the inflow of irrigation waters) their corn and pumpkin plants and she whispered to me as we got to a safe distance, 'you have to be ready to take the water, otherwise the next in line, your neighbour [gesturing to the men uphill from us] will take it'.[12] A *propietario* might arrive early, taking the opportunity to gather green fodder for their animals by pulling weeds, may *aporcar*, or take advantage of an imminent water source to safely burn off troublesome foliage and weeds encroaching on the *melgas*. As your neighbour tells you 'ya', you know you can take the water, realign the *compuerta* (sluice gate) so that they redirect the flow, '*echar*' (throw) the water your way and stand and watch for the *melga* to fill to the '*puntita*' (that is, be adequately filled). Waiting for it to do so, young people sit and check their phones, old people undertake small chores, or respond to this anthropologist asking things like 'has there been any more water since that big rainy season earlier this year?' 'The irrigation water in Peine is always the same', came many an answer. However, in Tilomonte, it can vary more. Each landowner receives a '*turno*' for each section of their fields, taking their turn every ten or so days. In Peine, there is enough water to irrigate fields for around half the day with the flow from the reservoir if it has a full head behind it, filling the first of the day's fields in a bubbling rush, and the last of the day's water taking longer to fill each *melga*. The quality of this flow is somewhat dependent on the quality of the *juez*, or the last person who closes the reservoir, with the timing affecting the head of water for the morning's irrigation flows.

The role of a *juez* is temporally demanding, contrasting with the flexibility of most other agricultural tasks. The *juez* must be *there*, they must ensure no-one takes more than their share and each *propeitario* who plants their fields takes only sufficient water. Not taking one's turn at the correct time makes the water system difficult to account for and can attract a fine, though some in the role of *juez* may '*cobrar*' (charge a fee) to irrigate fields in an owner's absence, as Justo explained to me, rubbing his forefingers and thumb together and grinning. Older men and women take on the bulk of this work, like all other agricultural work in Peine, some irrigating for those younger relatives absent from the task due to salaried employment, but not willing to leave their fields fallow. As Barthel (1986) noted in relation to this role in neighbouring Socaire, the *juez de agua* is a legal as well as economic one, because the *juez* has the power to enforce the local norms of agricultural water distribution. In Peine, it is each landowner who judges when their *melga* is full and indicates to their neighbour to open the sluice gates in their favour. While Barthel reported that in Socaire the role was alternated between two knowledgeable (and stern) men, in Peine, the role of the *juez* circulates. In normative terms, each landowner and irrigator takes their turn to spend a portion of their days in this labour during one or the next irrigation season, during which they are expected to make sure the water is managed and taken well. The number of days that a person must act as *juez* will vary from 10 to 40 days, a period relative to the quantity of land a *propietario* irrigates.

As festive and productive labour, the annual irrigation canal cleaning (*limpia de canales*) is a much-anticipated annual event held on different dates in the different communities of the Salar de Atacama. As a 'spring festival' (Mostny 1954; Barthel 1986; Grebe 2004), it occurs around the beginning of the irrigation months (September–October). I sought authorization from Peine's *asamblea* to participate on three separate occasions (in 2011, 2016 and 2017). It stretches for multiple days and is punctuated by libations to the water and ancestor-mountains along the route of work culminating in a final collective ritual, the Talatur. Materially, the multi-day event has the effect of leaving the irrigation ditches weed and sludge-free, so that the water flows better through them and reaches the agricultural fields with less filtration. The labour also performs human relations with the non-human mountain entities, the peaks that send snow melt as water to village springs, sustaining life.

In the years I was present for the canal cleaning, the days of labour were led by Don Gerardo, who holds the agricultural plans of the village. He not only intervened in mild disputes about which sector of canals marked the place where a *propietario* must join the work party, but also advised on fines for worker non-appearance or for untidy sections of canals where these entered certain landowners' fields, and at the designated sectors in which workers to rest. At rest stops, allocated according to '*donde los abuelos siempre detienen*' (where the grandfathers always stopped),[13] he poured *aloja*, or *ckilapana* in Kunza (a brew made by fermenting *algarrobo* tree seeds), for each worker, slowly making his way among the groups. I mimicked each worker before me, first offering a little of the *aloja* to the irrigation canals, drinking in a large gulp, then tossing the remains in the air toward the peaks, shouting '*vito vito*' as others joined in the cheer for the success of the work. In the final collective ritual of the Talatur, formal ceremonial offerings include the previous year's corn bounty mixed in with potent and life affirming coca leaf, *aloja*, and *parina* feathers (Kalazich 2013; M. Núñez 2015).[14] Thus, I learned about water and these ethics of its relation to people in Peine from working in the canals and participating in the public ceremony (2011), following the workers and assisting in serving drinks (as *tajniri*), preparing the communal meal and observing the less public aspects of ceremony (2016) and working intermittently in different stages of the canal and reservoir cleaning work and ceremonial activity (2017) in both Peine and Tilomonte. During the days of irrigation canal cleaning, I was instructed to work happily, work hard and cooperate freely with one's neighbours.[15]

Each of the named sectors of fields in Peine are fed by one of three principal branches of irrigation canals. Cleaning the 'San Antonio' reservoir on the second day of the *limpia de canales* in 2011, its mud revealed the dynamic of communal and personal, rights and responsibilities, as these relate to specific physical sectors of the agricultural lands. San Antonio reservoir had a floor of earth and rock, which over the course of each year accumulated thick mud and algae. The heavy work of shovelling this muck and flinging it up and out over the edge of the reservoir was done mostly by men. Before they began, the mud was measured laterally with string (Figure 1.3), with each measure proportionate to a named *propietorio*'s landholdings in the sector fed by that reservoir. Then, as names were called and *propietorios* or their nominated worker entered the tank, they became responsible for their portion of mud. Once all entered, workers began shovelling, flinging mud and algae up and out of the reservoir, sometimes aiming at the captain of works or women labourers clearing weeds around the edge of the reservoir (Figure 1.4).

On the afternoon of the final day of canal cleaning, the *cantales* take clutches of *parina* (flamingo) feathers that have been contributed by each landowner, and place these around where the springs bubble up before they are channeled into the irrigation system. Meanwhile the work party makes its way up the final reaches of the canals with jovial gusto. A portion of food from each household saucepan that contributed to the worker's lunch was provided to the *cantales* who added this to the oblation at the springs.[16] At the culmination of three days of labour, all gathered for the Talatur above the *estanque chico* (small reservoir). The

Figure 1.3. Measuring and apportioning the San Antonio reservoir, 2011.
© Sally Babidge.

cantales call out the names and gestured to each mountain peak, naming each spring, and encouraging these beings to provide abundance in the year to come. Workers hold one anothers' shoulders and dance in a tight circle around the *cantales* and captain, their instruments circling one way and then the other, singing the Talatur verses, whose tune is the water's own song.[17] Workers may also place a loop of cane around another person ('*ramear*') to signal their intention to '*jugar con el agua*' (lit. play with the water), and enter the *estanque chico* in pairs to bathe. When I participated in 2016, many young men adorned themselves with these loops during the afternoon's work and spoke loudly about who they might choose. They entered the tank, whooping and being '*alegre*', materially mixing their playfulness, their bodies and potential fecundity with the water, the reservoir, the canals, and by association, the fields.[18]

The *limpia de canales* might be considered an extended '*convido*'. The term refers to a broad spectrum of private, public, large and small rituals in which people express relations with non-human worlds. The Spanish verb *convidar* may be translated as 'to invite another to accompany' one or to share what one has (but not to give over fully) (Asociación de Academias de la Lengua Española 2023).[19] In Peine, a *convido* gathers

Figure 1.4. Cleaning San Antonio reservoir, 2011. © Sally Babidge.

human and non-human participants in a time and space, marking some project of human labour with hopeful invitations to powerful non-human entities to accompany it. In public *convido*, a *cantal* and other participants entice the Pachamama (sometimes *Pata Hoiri*, sometimes translated as Earth Mother, or Earth Being, and primary Andean non-human being), mountains and ancestors to their side in the earthly event. In the human act of sharing (with some *convido* referred to as '*pago*' or payments), people offer territorial non-human entities a bounty of attractive consumables such as alcohol, plentiful coca leaf, sometimes food, tobacco and adoring and encouraging words. For those who lead conversations with the sky, waters and earth, a named peak (*cerro*) is animate, may be referred to as '*tata mallku*' (great grandfather mountain), may communicate or give signs and be tutelary in the sense of a teacher or an elder. Water and water labour are a particular focus of *convido*, as irrigation practices induce scarce waters to feed arid lands with communal and cooperative management, and ceremonial activity is part of the technical production of a responsive landscape (Valenzuela and Moyano 2021; cf. Gelles 2002; Boelens and Gelles 2005). *Convido* rituals, such as at the stadium opening, or performed as part of agricultural and water infrastructure work, mark the contingencies between human labour and the proclivities of earths and waters in their support or otherwise of human constructive and productive activity. As it was explained to me by Don Bartolo in Peine, during the irrigation canal cleaning work and ceremony, as well as other *convido*, people say to the sky *'que caiga lluvias, que [los vertientes] haiga las aguas'*. He used the 'subjunctive voice' in Spanish to express uncertainty and hope, that the rain will fall, that in the springs may be water.

In 2016, the leaders, supported by members of the agricultural committee, won a project from the government agency INDAP to heighten the walls of the San Antonio reservoir and seal its floor with concrete. The concrete alleviates many hours of labour in the *limpia de canales*. While everyone I spoke to was pleased with the prospect of lighter work that year, it also prompted reflection. What would happen if the community succeeded in turning all the irrigation ditches into concrete canals? There would be less need for cleaning, and perhaps the accompanying communal work party and customary activity. Gathered for tea one evening with women in their 30s and 40s, Paloma (the only 'young person' on the agriculatural committee responsible for the work), raised the topic of canal improvements and reflected sadly on the concrete. She asked rhetorically, 'will this be the end of the *limpia de canales*?' Luisa countered that, no, surely they would simply clean the canals of rubbish, and the weeds that surround the canals would always need to be cleared back away from the water flows.

Irrigation water practice and relations among community members change as generations of landowners, individual and collaborative labour shifts over time and is negotiated, never assured. Irrigation water management is thus an achievement of collective effort, directed by the principal that Trawick (2001: 367ff) refers to as proportionality.[20] An ethic of irrigation water labour seems to be to maintain a sense of ease in relation to the communal title to territory and waters of the springs that feed the village reservoirs, by dividing, apportioning and monitoring each *propietorio*'s effort according to a local ethic of equity and productive gain. In turn, this labour and the good mood by which it should be undertaken is an inducement of the non-human others who dwell in the earth and landscape (Pachamama, peaks and ancestors) to join in ensuring the future of crops and animals and human life. Proper labour and effort are the responsibility of members of the community, and share a dynamic that partly depends on the character and efforts of those who perform the role of the *juez de agua*, take up a leadership role in the agricultural committee, or the temporary position of captain in the *limpia de canales*. While *propeitarios* respond to events according to a sense of proper worldly relations to non-humans in their worlds and negotiate relations among *socios* so as to continually seek to achieve community through water labour, there is also the strong character of particular persons who lead 'community action'. In addition to this kind of historical dialectic of the person and the community, a unique character of people in Peine is narrated, especially by older people.

A Short History of Local Ideas of Development

There is a story about a small plane crash near the village, that, when I first heard it, reminded me of the very short span of time between, on the one hand Peine being a 'very remote and poor place', as Doña Claudia (originally from Socaire) told me, and, on the other, the place today considered wealthy by other locals of the Salar.[21] It is a story that in some tellings has a hint of millenarianism, considering how a local fantasy of modernity may be tied to infrastructure and transport (cf. Harvey and Knox 2014). Below a retelling, rendered from conversations with various people in Peine. It is set in an 'out-of-the-way' place (Tsing 1993), sometime in the mid-century, perhaps late 1950s or early 1960s, before there was a road to the village, only mule tracks.

> A small plane appeared from across the mountains and circled above the village, apparently in distress. People gathered on the playa at the edge of the Salar and waved handkerchiefs to indicate that the aircraft should land in that open space. The plane eventually crash-landed in sandy ground near what is now the mining camp 'Pedoz' (owned by SQM), about 15km north of Peine. The pilot survived, but the plane had a cargo of meat (from Argentina) that could not be transported without spoiling. The authorities, who eventually arrived from San Pedro, allowed locals from Peine to take what meat they could carry. Over time, people from Peine salvaged parts of the plane for construction materials. Sections of the plane's fuselage remain visible, reminders of the historic event. Showing me around her garden the first time, Paloma smiled pointedly at a sheet of metal that had been used to fence off one part of a plot at the rear of her mother's house. Doña Inéz also told me the plane story when we were waiting for her sheep to graze in the grass of the sector of agricultural lands where the irrigation canal it had been improved using a part of plane body. She laughed at the anecdote as she told it and indicated the ingenuity of that older generation (see Figure 1.5).

Talking about this with Ivana, she reflected on what she had been told by older people, who considered the plane crash an important sign of things to come and welcomed the potential of more outsiders reaching the village. They organised communal work, flattening and clearing a landing strip at the edge of the Salar, close to the community. Other planes might

Figure 1.5. Part of an irrigation canal extended using a section of plane fuselage. © Sally Babidge.

thereby land safely, perhaps even plan to come. Ivana said, 'they [meaning the older generation] were always thinking of bringing outsiders to Peine'. The plane crash sparks reflection on the potential of attracting wealth as part of a local history told about the self as well as the kind of cultural observation recorded by others.

Located in a frontier zone, and historically part of a shifting national border and resource territory, Peine has been 'remote' from industrial modernity. Atacameños have been subject to outsiders judging them as having questionable capacity for modernity, like many Indigenous, peasant and rural peoples. San Román (1896) and Bertrand (1885) were Chileans reporting on the interior of the Antofagasta region, with a special interest in considering the extent to which Atacameño peoples might be 'able' to become Chilean, given their 'barbarism' (in Mercado 2007).[22] Atacameño peoples were important to the Chilean state's sense of the desert 'mining' territories as part of a rich but tenuous frontier (Gundermann 2002), just as the Bolivian state had been aware of the Indigenous peoples' propensity to take sides against colonizing forces (Sanhueza 2001). For example, San Roman finds a schoolhouse in the 'small hamlet of Peine' in which men and children were writing in 'elegant calligraphy'. It provides a 'marvellous portent at this altitude', he reports (San Román, in Mercado 2007: 11), that is, a sign of their capacity to become citizens. The evident interest of people to speak Spanish (P. Núñez 2006), and the fact that the community in Peine had organized a schoolteacher, was considered a sign of 'civilization' relative to other Indigenous peoples the writers encountered in their travels in the region.

A local sense of self parallels these reports, though the tone and emphasis differ. Don Lorenzo, a man in his 80s, agreed to talk to me about his experience of social change in Peine during his lifetime. When asked about how he understands the community's relationship with mining companies, he answered indirectly, making a comparison to neighbouring communities, and rendering his image of Peine in terms of its character.

> How do I say it? People in [a neighbouring village] were more Indigenous.[23] [In Peine] we were more *despierta* [awake] than in other pueblos, people invented things and were very *cariñosa* [friendly], they would get together and do things . . . work together. The things we promoted were things like national holidays, we respected the church, respected everything like that . . .

In Don Lorenzo's mind, the other village focused only on producing agricultural products; in contrast, Peine was outward-looking, taking on Chilean National values. His meaning in answering my question was implied, which was consistent with the way his community had responded to the mining industry. This character, and the history of the community's worldly interests, is also reflected in older men and women's stories of the beginning of the lithium mine (see Chapter 5), in which '*todos los cabros*' (all the young men) leapt at the chance to gain work on the Salar when the lithium mine opened. With more local work becoming available, people enthusiastically returned to Peine, their families and ancestral homes, then, as they are doing now with the lithium boom.

In the early 1940s, Peine had a schoolteacher who came from Tocopilla as a young man, married a local woman and stayed for more than twenty years. The man is remembered in Peine today for his violent punishments of school children who participated in customary activities (such as the irrigation ditch cleaning) or spoke the Kunza language. The teacher's presence for some decades thereby directly and forcefully intervened in the intergenerational transmission of language and lore (see M. Núñez 1998: 196 ff; Kalazich 2013). However, in Núñez's work of the late 1990s, her interviews with Peineños demonstrate their interest in both lamenting this linguistic and 'cultural' loss and expressing pride in and a value for the modernity of the village and its young people. By the 1990s, a good handful had professional training and some had begun to reinvigorate a communal interest in customary practice and traditional cultural knowledge.[24] A more detailed historical ethnography of Atacameño-Likanantay modernity may be warranted, but it is not my intention. I want to emphasize two aspects of the story: people I spoke to emphasized that the village itself had insisted on having their own schoolteacher (unlike other places) and visions of development are central to the ways in which people in Peine see themselves and their mode of participating in what the world has to offer.

Development, as it is applied in common speech by my interlocutors in Peine, and the company employees who are responsible for company-funded development projects, is moral thought applied to socioeconomic change associated with modernity. Its local usage refers to negotiations among *socios* and *propietorios* about locally envisaged change and their equivocations. It is the same term used by state and corporate actors when they imagine the future economic 'good'. As in any community, among Peine's community members are many different approaches to thinking about, responding to and working in the extractive industry as associated with 'development'. Diverse perspectives on the socioecological and economic effects of the mining industry are themselves contingent on internal dynamics as I witnessed in the contexts above and others. These perspectives arise in and from this place and its social and ethical terrain, remote from the state and the centres of power, yet central to so many global dynamics. Since my first fieldwork visit, I have witnessed a range of different views rise to the surface through the election of community leaders and the actions they take. At this time or some other, what does 'the community's vision' consist of? At times this temporally contingent representation is driven by community representatives who desire agreements for 'development benefits' with mining companies. At other times what has emerged is a set of elected leaders and speakers for the community who, relying on their knowledge of the region and their sense of relations at stake, have sought to outright resist mining activity or its expansion.

Peineño response to the conditions of advanced extraction as a 'slow resistance' consists of both their ongoing engagement in the systems of advanced extractivism and an incomplete acquiescence to these systems through continuing to insist, for example, on the *asamblea*'s role as the decision-making locale. This contrasts with commentators such as Rivera (2004), who interprets Atacameño engagement with state agencies and mining companies in terms of a parallel between the 'contractor logic' of extractivist interests (the reduction of the large mining companies to bands of contractors) and the same logic emerging in the practices of communities. It is an interesting interpretation regarding the hegemony of neoliberal systems for bidding for funding and their influence. Such practice is part of everyday realities for communities and others, but the idea that the neoliberal logic of 'hyperflexibilization' of life is the overriding logic of everyday Atacameño practice is not sufficient for the diversity of local orientation I have encountered. While structural relations of neoliberal conditions are indeed dominant, they are exceeded by how people make daily judgements about what may be good for the present and future. The differentiation matters. Instead of assuming that engagement in capitalist practice defines the logic of the person engaged, following Bear (et al. 2015) we may ask, 'what makes capitalism in one or another place seem coherent'? That is, how does people's engagement in the dominant industrial economy make it seem like they are simply subject to these logics and what does this miss about people's interest in producing a future otherwise?

The following chapters examine in ethnographic detail the ways in which extractive capitalism is far from an all-encompassing logic in the lives of people in Peine, as well as how claims to capitalist mastery make life difficult for extractive company employees. I return briefly to the vignette of the Llullaillaco stadium opening to further consider the frameworks of company operations within advanced extractivism.

Ecologies of Late Capitalist Development

At the stadium opening, Peine community members welcomed the chance for a day of festivities, and leaders articulated relations with the mining company with no great animosity, despite muted and public conflict that existed between them. There was a show of agreement, though also many signs of disavowal. The pre-event *convido* at the back of the stadium included a few junior company employees, including an ex-employee who told me pointedly that he 'no longer works for BHP', and all were welcomed by the organizers to participate in the invocation of vital relations between Atacameño people, water and mountains (with special mention of the Llullaillaco peak). None of the senior company people were present. Also absent were many community members. While the FME Director, was emphatic about 'convergence', 'joint work' and a collective being 'part of nature' which can be 'made the most of', he surely knew very well that the relations between mining companies and communities in the Salar de Atacama, which go back to the 1980s and forward into the future, have never reached and are unlikely to gain any such harmonious distinction. Instead, we may see that his statement was an advanced form of the rationality of extraction as development which 'recognizes' different worlds to the extent that this recognition can be put to work to sustain extraction.

As expressed in events like the *convido* and the *limpia de canales*, Atacameño-Likanantay peoples in Peine described relations with the waters and mountains of their territory in terms of the earth itself as a being. In this register, earths and waters are animated *'seres sensibles'*, sentient beings, with whom one seeks to maintain relations. This form of relating, living and thinking is like that of long-standing trading partners across the region, some also identifying as Atacameño (including some in present-day Argentina and Bolivia). It is also familiar to anthropologists and others who have described Atacameño-Likanantay ecological practice in detail (e.g. M. Núñez 2015; Castro and Aldunate 2003). And these 'Andean' practices articulate with the cultural and hydrological relations to non-human beings in the world in ways well known in the discourse of regionality debated among anthropologists (Starn 1994). A broader recognition of culturally meaningful environments and landscapes, perhaps rendered in terms of 'cosmology', are part of the indigenist policies of government and those used by civil society to defend people and ecologies in the fact of extractive impacts (Canessa 2005; De la Cadena 2010; F. Li. 2015).

Andean perspectives on landscapes have also become part of the discourse of corporate or 'extractive-led development', such that mining company managers and social relations employees might assert that these matters, as beliefs, can be accommodated by company policy. The way in which the FME Director spoke of Andean ecological relations at the stadium opening, that – 'we are all part of nature' and 'can take advantage of it' – enacts an approach that is a kind of extension to what Escobar (1995) referred to as 'environmental managerialism' in Latin American development. The environment in development is a form of instrumental nature allowing humans to draw upon it as a resource for human well-being and manage ecological relations with scientific mastery. According to Escobar, environmental managerialism endangers diverse peoples and places because it asserts a singularity of 'the Earth' in a universalizing 'western science' (1995: 193–4).[25] As 'management', the threat is as mundane as it is encompassing, an everyday assumption about how things might best function.

The late capitalist expectation that diverse ecologies, relationships and social actors can be managed finds an example in ambitious projects such as the United Nations' Millenium Ecosystem Assessment (MEA). The project of planetary assessment sprang up at the turn of the twenty-first century; 'From 2001–2005 . . . [the MEA produced] a state of the art scientific appraisal of the condition and trends in the world's ecosystems and the services they provide, as well as the scientific basis for action to conserve and use them sustainably'.[26] The MEA marks a moment in which global capital and developmentalism advanced from 'environmental management' to 'ecosystems' thinking in seeking to address the planetary questions of the Anthropocene. To quote the Millenium Assessment authors' conceptual definition of 'ecosystem':

> Humanity has always depended on the services provided by the biosphere and its ecosystems. Further, the biosphere is itself the product of life on Earth. The composition of the atmosphere and soil, the cycling of elements through air and waterways, and many other *ecological assets* are all the result of living processes—and all are maintained and replenished by living ecosystems. The human species, while buffered against environmental immediacies by culture and technology, is ultimately fully dependent on the flow of ecosystem services. (italics mine)

The core concept of 'ecosystem services' in the MEA aims at the systematic integration of ecological and economic values; how do earths, waters, animals and plants each and together in certain regions, and as part of a planetary unity, provide 'services' to humans' 'wellbeing'? (cf. Marzec 2018) As the definition makes clear, this form of thinking posits a consensus, led by globalizing civil society, that the problems of the Anthropocene can be addressed according to a 'unified planetary ecosystem' (cf. Fortun 2012). As with the other Millenium Assessment case studies, the report proposed to account for a regional 'ecosystem' (RIDES 2005). Its analysis acknowledges the assistance of an advisory committee whose 18 members included the president of the Atacameño People's Council, two other Atacameño residents of San Pedro de Atacama, four representatives of the four principal mining operations in the Salar and the remaining members from state agencies or local government. Minera Escondida, Zaldívar and lithium miner SQM contributed 'financial and in-kind support to the research'. The 'stakeholder diversity' of the list and accompanying photographs of workshops make testament to an apparent inclusivity of the research process. Unsurprisingly, the authors found that water 'is the central issue' to the whole area and to all human actors; 'All other activities (goods and services *provided by the Salar*) – mining, biodiversity, agriculture, etc. – are directly related to water quantity and quality' (from website, ibid., italics mine).

Some years after it was published, I read the RIDES MEA report and the separate report titled 'Patta hoiri and Likanantay people', which refers to interviews with locals undertaken in 2004 (Bustos and Blanco, n.d.). While the main report and its 50-page 'Executive summary' provides a smooth scientific image of assessment in line with pretension to a globally systematic reporting, the separate 'cultural' report provided an overview to Atacameño-Likanantay community, history and ecological life. The authors refer to 'special characteristics' of the region that researchers encountered, partly as an explanation of how their work and timely assessment was hampered. They wrote,

> Efforts were made to disseminate the project amongst the community and municipal authorities, in order to encourage participation from diverse stakeholders. This process met with resistance . . .The community has developed hostility towards initiatives proposed by foreigners and foreign institutions, due to both saturation by numerous cooperation projects and meetings and previous negative experience [sic] where the results of investigations and projects were never shared with them, and the response to this project proved no exception. (Bustos and Blanco, n.d.: 1)

Approached by MEA researchers for a proposed case study in the Salar de Atacama, Indigenous peoples questioned the value of the study and the nature of its research model. The separate report instead set out local testimony, interviews and published material on Likanantay-Atacameño knowledge, practice and science, and detailed then-current movements to protect and revive Indigenous knowledge. Researchers encountered people's insistence on an ecological relation with the Salar and its human and non-human ecologies, and how these relations are practiced. As a separate report, unable to be integrated into the MEA's system, it stands testament to the incommensurability of the ecologies of the MEA and local people.

In the main report, the authors concluded that residents of the Salar had concerns about inequalities in access to water rights, especially between industrial and community users, but nonetheless 'water consumption by mining does not affect or interfere directly with other consumers in the Salar de Atacama basin' (RIDES 2005). These findings parallel company reporting from this time period, and as I show in this book, have been sustained over the many years following. One may conclude that the MEA, as globalizing environmental governance, supported the industry's capacity not only to continue extraction, but to *expand* their extractive activities. The conceptual framework and definition of this assessment describes the market economics that drive these approaches and helps explain how such an assessment can decline to identify the extractive industries and industrialization more broadly as key actors in actual activities that cause ecological harms. MEA's conceptual framework report makes this clear:

> [The MEA was initiated] in recognition of the growing *burden that degraded ecosystems are placing on human well-being and economic development* and the opportunity that better managed ecosystems provide for meeting the goals of poverty eradication and sustainable development . . . (Conceptual Framework, Millenium

Assessment, https://www.millenniumassessment.org/documents/document.765.aspx.pdf. Italics mine).

If an ecosystem is animate, 'providing services', it is also threatening (withdrawing said services), since 'degraded ecosystems' fail to sustain humans. This perverse animism of 'ecosystems' imagines ecological things having only instrumental ends (serving or declining to serve). While advocating the study of ecologies as interrelated biota, the MEA accounts for the value of ecologies according to their ultimate sustenance of humans and human activity – both managing human attitudes and uses of 'ecosystems', and human economic potential as derived from ecosystems. This is a 'veritable ontological occupation of relational worlds' (Escobar 2017: 1). Such capitalist animism (Taussig 1997) has become part of the play of the material and political nature of extractive contests, while ecologies and their contingent relations with local peoples continue to intrude on the neat telos of these projects.

The MEA conceives of the possibility of a systematic and scientifically informed governance that incorporates scientific accounts of ecological systems into economic accounting to secure business as usual. However, as the separate and particularistic report by the RIDES authors for the Salar de Atacama case study made clear, Likanantay peoples sometimes insist that scientific research ('by foreigners and foreign institutions') and ecological management in corporate terms does not serve their interests. The RIDES report also makes evident that a negotiated relation with water is exactly what draws mining companies and communities into a relation with one other, yet their relations are always agonistic, if not antagonistic.

In the many descriptions that I have read in reports authored by state, corporate and global civil society actors, of the waters of the Salar de Atacama basin and how these waters arrive in the Salar, matters of relation between Likanantay people and non-human things and beings are either absent or represented in a separate, cultural mode. Ecological relations like those described in the separate RIDES report, and those negotiated in *convido* rituals like the one performed at the stadium opening, or those in the *limpia de canales*, articulate and perform dimensions of both kinds of reality. On the surface, this may be explained or recognized as the importance of nature or the ecology to Indigenous people, which can be managed by (say) a separate report about local diversity of knowledge. In the separation of these concerns (a separate report, a separate ritual), it is a clear expression of the problems of incommensurability between the ecologies of universalizing rationality of advanced extractive capitalism and the people's ethical commitments to one another and their world.

The classical challenge from anthropology to the equivalence made in universalisms is that while some equivalences sometimes work very well, different people assign different values to the same things. The universalizing rationality of equivalences is reminiscent of Hegel's 'cunning of reason' (1953), as Sahlins (1993: 12) noted in parallel observations of the perverse application of culture in Western capitalism; 'a cunning of culture which, by harnessing an absolute sense of rationality to a relative logic of signs, ushers in a (truly) golden age of symbolic freedom'. The notion of cultural relativity, warmly embraced by late capitalism, is a kind of freedom to assign different values to different things, allowing for the diversity of perceptions of the world to be encompassed by rational thought. State (and corporate) recognition of Indigenous peoples and rights according to liberal thought may be analysed in the same way, as Povinelli (2002: 4) observed in relation to the recognition of Indigenous peoples in settler-colonial Australia. However, the rational knowing about Indigenous difference under regimes of recognition (Indigenous peoples are sufficiently recognizable to hold a delimited set of rights) produces 'irresolvable obligations' embedded in 'deontological demands' made of Indigenous peoples, while the incommensurate aspects of Indigenous worlds (a full life) are excluded.

In parallel, extractivist claims to know about 'ecological' belonging and an Andean concept of the good life (*el buen vivir*) are claims to liberal values and capitalist mastery; it is only rational to see how important nature is. This dominant regime of knowing that Descola (2013) called 'naturalism' retains the distinction between nature and culture and misses the 'excess' of incommensurability (de la Cadena 2015) between the worlds of the extraction company and those of Indigenous peoples. These are differences not encompassed by liberal recognition of cultural difference and the limits set aside the material facts to which people are committed and which are understood to animate different parts of their worlds.[27] The potency of non-human worlds, the peaks as beings and their attendance of water sources such as pools and springs, are present in human activity, even or perhaps especially so, in response to the imposition, threat or promise of global capitalism and mining.

Conclusion

Relational commitments to such non-human entities that co-inhabit the world with humans are expressed among other Andean peoples (e.g. Allen 2002 [1988]; Castro and Aldunate 2003), landscape and other vital ecological agents are structured according to an interpretable logic (they are related and known), but they are also characterized by the potential for unforeseen irruptions (Göbel 2001, 2008). That there may be an inherent uncertainty in all subject relations is also how I understood the world of relations according to people in Peine. Factors of uncertainty are myriad, but in relation to water, for example, the variable quality of spring waters demand experience, though a person's own history and knowledge of the origin of certain waters may be respected within limits, because waters change. Water's history and its geolocation are uncertain, not only, but also due to the impact of human activities, their own labour or climatic factors that affect the conditions of flow. Water (as 'nature') does not constitute one combination of life and non-life, an environment, that each human actor in the community or those from mining companies share. Water may be treated as an 'asset', an objectified resource of whose value community members, companies (and their employees) and the parties to a company-community agreement may 'take advantage'. Water is also celebrated in ritual as a being, and its material flows generate life (*'que sea en buena hora'*, all things being well, in the common phrase uttered at such times). But, as outlined in the work of de la Cadena (2015), such descriptions are 'not only' how such relations are, and I argue that they are also not always how relations are made and become.

The MEA report I describe briefly and the FME Director's statements to me at the stadium opening are examples of late capitalist attempts to reconcile ecology with economy, subsuming other values and relations in the process of harnessing the agentive aspects of earths and living beings for capitalist 'development'. This rationality activates the value of ecological subjects as resources within a complex but manageable 'ecosystem'. Corporate environmental management reporting, and the idea of ecosystem services present a mastery of the world in its regional scale (relative to particular mining operations), or more planetary forms. Luisetti refers to such rationality as 'geopower' (2018; cf. Povinelli 2016), a kind of planetary ecological management by which business-as-usual may prevail. It does so in the register of a capitalist animism, discursively bringing order to the uncertainties of ecological subjects. This is advanced extractivism in operation, comprised of the practical and moral dimensions of an accounting of 'ecosystem' activities (the generative nature of nature) as part of the recognized rights of Indigenous communities and responsibilities of companies.

In advanced extractivism, ecological objects are delimited to manage contingency. Other matters of social or ecological concern may be integrated into such a masterful accounting system if something can become a value in the model and help to justify and sustain the ongoing conditions of extraction. In his comparative ethnography of extractive capitalism in the Pacific and Latin America, Kirsch (2014) observed that the critical reports by Indigenous activists and their allies about the social and environmental harm caused by mining are incorporated by extractive capital into revised policies of 'best practice' in social and environmental management. A key critique of extractivist systems, Kirsch's work demonstrates the reasoned effectiveness of developmentalist thought in the operation of extractive capital. As the insistence on participation in the MEA process according to their own terms (yet quarantined in a separate report) made clear some years ago, the apparent mastery and dominance of extractivism in the Salar de Atacama is not the whole story. So too in Peine: neither industry-alignment nor a wholly resistant stance characterize how people in Peine respond to extractive capital.

In the community leaders' speeches during the stadium opening main event, they emphasized community members' labour in completion of the stadium, an achievement they explicitly traced to their own or another's good community leadership. Each speaker only acknowledged the financial support of the company in passing, and one noted that the stadium had been finished properly, despite conflict with the company. This may have been a veiled reference to the impending negotiations associated with the question of Escondida's groundwater extraction. Nonetheless, I also interpret their practiced emphasis on community interests as a dimension of 'slow resistance'; an interest in the primacy of community relations and the value of community labour at the same time as a refusal to fully accept the terms of corporate recognition of 'their interests'. The ways in which Peineños work with one another and with water narrates something of how people's sense of uncertainty and contingency is locally managed and how such indeterminacies converge with advanced extractivism.

Notes

1. *Escondida*, the corporate literature says, refers to the principal ore body's position, which lay beneath hundreds of meters of 'overburden' or non-ore carrying earths.
2. Figures are an estimate based on analysis of the public register of the Water Authority (Dirección General de Aguas, DGA). See Chapter 2.
3. Minera Escondida is one of a handful of mega industrial-technical sites of copper extraction in Chile's Atacama desert operated by BHP.
4. All quotations from interlocutors were originally in Spanish and the translations are mine. I have left particular words or phrases in Spanish where it is particularly relevant to my interpretation, or the translation risks the loss of meaning.
5. They may not make a great team, as Presidents have explained to me at different times, you have to work with the *socios* who are elected, and sometimes the *directiva* is composed of *socios* with very different experiences and perspectives, or understandings of their social, economic and political worlds.

6. This number includes meetings as *asamblea* of the Atacameño Community of Peine and *asamblea* of the Junta de Vecinos, sometimes with visiting state agencies or mining companies. Over thirty meetings also with the leaders of the Community or JV (between 2010 and 2019).
7. This was never paid work, rather, an opportunity to reciprocate for the privilege of the research relationship.
8. Except, of course the quote from the president in this paragraph. With the president's knowledge, I visited with people in their homes and checked permissions in regard to their personal intellectual property that is included in these pages. Also, employees of the community's Environment Unit requested my assistance in providing background information in a case against a mining company, and I produced a short report to the president and *directiva* on that matter.
9. The original in Spanish: '*la asamblea manda. No deben tomar decisiones solos*'.
10. Discussed further in Chapter 3.
11. For example, in 2011, before she passed land on to one son, and later took up the cultivation of another set of *melgas* from an aged cousin with no heirs, Inés had three small *melgas* in the lowest sector of Peine's fields, four above the San Antonio reservoir, two adjoining her elder sister's in the upper sector and two a little higher, as well as *melgas* in three different sectors of the agricultural oasis Tilomonte, 15 km south of Peine.
12. I was sitting with a gathering of older women that same year, who joked and complained, making exaggerated swiping motions on imaginary hand-held phones, about not liking these new computer telephones, since all the young people publish everything about their whole lives including images of them, but these telephones are good for letting your *vecino* know, or finding out yourself, where the water is.
13. In 2016, on the first day of the *limpia de canales*, a rest stop was called by one of the captains (elected leaders of the work for that year) just as some of the work party reached a section of deep shade. Don Gerardo insisted everyone return down the incline to the time-honoured rest stop.
14. See Gil Garcia (2011), whose comparative review of the Andean literature on rain-making ceremonies finds many similarities in these practices across the region. Kalazich's and Núñez's theses, both cited here, are expansive on matters of the *limpia de canales* and *Talatur* in Peine and neighbouring Socaire. These theses are available to readers of English and Spanish. My description is thus aimed not at discussing the detail, but emphasizing the joy and seriousness with which these labours and the ceremony is undertaken, as well as indicating how the annual ceremonial work party illustrates the fact of water's multiplicities, in this case, its relations as they are practiced in the ethnographic present of the research period that informed this book.
15. Cooperating with one's neighbours can be understood in terms of sharing the labour time during the *limpia de canales*, when a person is lined up to work in the canal alongside the person whose territory adjoins yours. Thus, for example, in 2011, when I wielded a shovel for Doña Inez's land on the final day of the events in Peine, I happily worked behind and alongside an older woman who worked her deceased brother's fields, which adjoin Doña Inez's fields in this uppermost sector of the *monte*.
16. See details in Núñez (2015: 243 ff), for matters of ceremonial and ritual relations between the more-than-human beings and people in Peine.
17. In addition to those already cited, discussions of Atacameño practice in Peine in these terms can be found, for example, in the work of Mostny (1954), Castro and Varela (1994) and Grebe (2004), Kalazich (2013) and Núñez (1998, 2015). The Talatur as the water's own

song was explained to me by those with whom I participated in singing it in 2011 and 2016, and also in a short film available by Gedda (2010).

18. See Mostny (1954), Barthel (1986) and Núñez (2015: 264).
19. It seems that there is no direct English equivalent. I use '*convido*' throughout.
20. Referring to research undertaken in three different Andean villages in Peru, Trawick identifies 'six principles governing irrigation practice: autonomy, uniformity, continguity, proportionality, regularity and transparency' (Trawick 2001: 367). The typology is less a concern than the broad parallels to my observations in Peine: people say that everyone *should* know how to run the whole system, each *propeitario* plays an active role in monitoring the whole, and the reciprocal rights ensured by each person fulfilling their duties maintains the sense of equity in access to irrigation water.
21. An Atacameño friend from one of the communities of San Pedro de Atacama reminisced with me about how, in her 20s, she expressed a desire to own her own *camionetta*. Her brother told her she should marry a Peineño.
22. These authors wrote at the time of the end of the War of the Pacific, when Chile gained the mineral rich northern territories, including the Salar de Atacama, from Bolivia.
23. His meaning was meant in pejorative terms, referring to his perspective of neighbouring peoples as having a greater interest in conserving tradition.
24. As Sahlins (1993) observed, and a number of anthropologists have since explored, an interest in the revival of Indigenous 'tradition' and professional training or successful business may strongly overlap, thus articulating the argument for the diversity of 'indigenization[s] of modernity' (20–21).
25. Escobar's recent work (2018) on the 'pluriverse' seeks to not only correct the assumption of universality of western science, but interrupt the ongoing project by proposing design theories for worlds not confined by such thinking.
26. From website: www.millenniumassessment.org/en: 'About'.
27. This is insufficiently detailed to claim to be an account of Likanantay ontological schema. Instead, I seek to indicate that ecological and economic material (or worlds), including the bodies of humans and the waters and earths to which they relate and on which they depend, are not entirely the same in practice or in reason to the bodies, earths and waters of industry actors.

Chapter 2

TILOPOZO

Ecological History and Extractivist Enigma

'There are more uncertainties than certainties' the facilitator told the meeting. It was January 2020, the *asamblea* in Peine's *sede* (meeting hall) numbered almost forty, and we were gathered for a crucial stage in the consultation process for Escondida's Environmental Impact Assessment (EIA) for the 'Monturaqui Project'. Escondida submitted the EIA in 2017, in support of its application to extend its license to extract freshwater for a further eleven years from part of an aquifer ('Monturaqui'), which is related to wetlands and salt lakes at the southern end of the Salar de Atacama. For over twenty years, Escondida's copper mining operations sustained growth in pit size and quantities of mineral processed thanks in part to the low financial cost of accessing groundwater to process ore. Their EIA arrived in the offices of the government Environment Evaluation Services ('Servicio de Evaluation Ambiental', SEA) and in line with regulations and in response to Indigenous demands to do so, the regional director of SEA ordered a period of consultation. Tilopozo is a 'protected' place and an ecological and economic subject in the history of Peine and in community members' visions of the future. The *vega,* or wetland pasture, is also a surface indicator of any environmental 'impact' associated with Escondida's and other companies' water extraction. Thus, Escondida's EIA needed to assure the public, especially Atacameño-Likanantay people, that if the company extraction license were extended for a little more than a decade, the aquifer would continue to nourish the social and ecological relations at the surface.

Over the following two and a half years, Likanantay communities of the Salar de Atacama, the regional Indigenous peak body (the Council of Atacameño Peoples, CPA), and other social and environmental actors challenged the claims made in the EIA, which led to the company reviewing

Notes for this chapter begin on page 82.

and modifying its proposals regarding future quantities of water extraction and measures of harm mitigation. At the end of 2019, Escondida's water rights in the Monturaqui sector expired with still no resolution to the company's application to extend extraction rights to the aquifer. A month later in Peine's *sede*, three government employees from the SEA Antofagasta regional head office were present. They had the job of explaining Escondida's proposals in the EIA to the assembled community members and outlining Escondida's amended proposal; providing 'impartial' technical information to the Indigenous community consultation process.[1]

At a long table at the front of the hall sat the president and directors of the Peine community next to their team of chosen professionals: an anthropologist, geographer, hydrologist and two junior lawyers. The SEA part funds these external professionals, but their first responsibility is to report to the Peine community on their evaluation of the company's scientific and social scientific claims in the EIA. The Peine community had also engaged the services of a nationally renowned Indigenous Rights lawyer, who had agreed to represent Peine's interests in this matter.[2] The afternoon meeting wore on as each professional presented their report, and *asamblea* members came and went, but by the final half hour, there were some seventy community members in the hall, all poised to hear the summation by their senior lawyer. She repeated what the meeting's facilitator had opened with, saying, 'there are more doubts [about the Monturaqui EIA and its implications] than certainties'. This meant that decisions about the environmental merits of this extraction project could not be taken by anyone, she insisted, including, and she gestured to SEA representatives in the room, the government authority. Given the level of uncertainty, she argued, the SEA had an obligation to stop Escondida from extracting more groundwater since there was no independent scientific evidence about extractive impact or sufficient company plans regarding the mitigation of that impact. Following her summation, Peine community members present were called to vote yes or no; were they prepared to go into negotiation with Escondida over their EIA? The question, she explained, meant that if they voted no, this would be taken to the SEA and count against Escondida's plans. If they voted yes, they would thereby agree to endure ongoing ecological and social uncertainties of the impact of Escondida's continued water extraction.

The material and political relations of groundwater to the people of Peine, the extraction industry's exploitation of groundwater in the southern Salar, and government regulators' activities each have a hand in the story of uncertainty. A growing literature on groundwater and groundwater depletion points to the problem of uncertainty and subsurface 'invisibility' (Birkenholtz 2015; Molle and Closas 2019; Gleeson et al. 2020;

Walsh 2022), as well as inherent indeterminacies in mapping and knowing aquifer systems (Ballestero 2019). Bessire (2022), for example, characterizes groundwater in terms of 'aporia', meaning the 'irresolvable internal contradictions, epistemic ruptures, and logical disjunctions within any given mode of perception, knowledge and power' (fn3). His characterization points to both a material and manufactured uncertainty in relation to groundwater. Its geophysical properties mean that groundwater is not visible on the surface, though springs and seepages tell stories about its lively presence and mobility, and theoretical models built on measurements of its different activity make it see-able. Since its subterranean unknowns began to be tapped using technologies that enabled pumping and drilling, groundwater has become almost half of all water used by humans and is increasingly threatened with pollution and exhaustion. In 2022, UNESCO declared the need to 'make the invisible visible' given the human planetary reliance on groundwater and increasing threats to its exhaustion. While recognizing the need for greater water management that UNESCO's proposals highlight, the case of groundwater in the Salar de Atacama shows that 'making visible' is a political as much as scientific technology. Emphasizing the invisibility and uncertainties that throw into doubt the capacity for effective regulation, as did the Community of Peine's lawyer, may be an effective political and ecological tactic in the interests of protecting groundwater and its relations to ongoing ecological and human life.

In the southern Salar de Atacama the fact of ecological harm from Escondida's (and other extractors') activity has been difficult to substantiate. The fact of uncertainty has enabled ecological harm to be both denied by the company and assumed by its critics, while aquifer extraction continued. This chapter begins by asking why Tilopozo matters. It examines the ecological and material histories and potential futures imagined by people from Peine in relation to the wetland pasture and spring called Tilopozo, a verdant surface surrounded by aridity, in the context of potential harm to it generated by the exploitation of groundwater. It then narrates the material facts and histories of the uncertainties of aquifers and environmental harms by considering both the hydro-ecologies and arid landscape of the surface of the southern Salar and the apparent subsurface capaciousness of its aquifers. Hydrogeological studies of the area, government regulatory mechanisms, company 'transparency' reporting and community participation in monitoring each trace different contours of uncertainty about the ecological subject (the aquifer) and the effects of extraction on it and Tilopozo.

Ecological Histories at Tilopozo

Tilopozo, abutted by crusty salt pan, is rare pasture in desert terrain and good water for humans and animals to drink and bathe in. It sparkles in local memories and imaginaries of a good life. Peine community members and Tilopozo share an ecological history as relations materially connected in their co-activities through time. People living alongside and working with Tilopozo's waters have channelled and cleaned the springs, burned and harvested pasture, drunk and dipped in the waters and let animals munch its grasses and fertilize new growth. Simply visiting, people may express their relations with a *convido*, an acknowledgement of the potential of any human action to be consequential for water and place.

When I visited the principal pool of Tilopozo with Doña Inés and her son one time, she kneeled at its edge and scattered coca leaf, also replenishing her cheekful, before using a stout stick to drag long green tendrils of algae, reeds and vines to the bank. The vine had an edible green fruit, she showed me, chewing on the '*nori*' (probably *Stuckenia filiformis*),[3] and sheep enjoy it as food variety. She threw some into her bundle of grasses and reeds (mostly '*unkillo*', (*Juncus balticus*)) that she, her son, and I had harvested and loaded onto the tray of the vehicle. We gazed into the newly cleared pool, and she indicated a point around two meters down at the bottom, where pebbles and sand gently swirled. Starting a comment to me in her familiar way, she said 'you know what, Sally? I was here when the Lascar Volcano erupted one time'. Rocks and smoke erupted in the distance, and the pebbles and sand in the pool whirled thicker and faster. Another time, an earthquake caused the water to move violently from side to side, Inés said 'like it was about to leap into the air', and her herd ran wildly to the southwest and higher ground.

This pasture area (*vega*) has 'always' been part of life for people from the locality of Peine. In the summer of 1853–54, the naturalist Rodolfo Philippi (1860) travelled by mule the 1,000 kilometres from Copiapó to San Pedro de Atacama and back, recording the quality of geological formations and biological life at and between *aguadas*, the infrequent and mostly brackish water sources along an inland route through the Atacama. Philippi refers to a man called Jose Maria Chaile, a *guanaco* hunter, pastoralist and artisanal miner from Peine. He met Philippi's small party when they stopped at Tilopozo to allow their pack animals to graze before the long journey south through a stretch of desert with less reliable waters. On colonial maps, Tilopozo is a rare place with good water that by its presence entices possibilities of exploration despite harsh desert frontiers (e.g. Mahlman 1854). The surface freshwater is, like springs all over the world, a sign of actual and potential life, a metaphor for wellbeing, as well as sensory evidence of what lies beneath. Extractive potential is also prefigured in mid-twentieth century US geological surveys, which found 'large quantities of ground water in the [Salar de Atacama] basin' (Dingman 1967: 2).

Regularly before, and into the 1970s, mainly during the winter months, women and men from Peine pastured their flocks of sheep and goats in Tilopozo. They might leave up to 200 animals to forage there when the highland pastures were covered in snow, though sheep, goats, mules and donkeys were grazed there at different times of the year, along with the saltier patches of pasture closer to Peine such as Palau and Cololau, to which Inés's sister and others took their herd. Inés remembered,

> We were very peaceful staying there . . . We would look after the sheep, take them from the corral and let them into the vega. And we would cook there, with the children. The children were happy, they bathed and played there. [One herder had] chickens there as well and they would get into the vega and lay their eggs.

A house was built for communal use at Tilopozo in the 1950s, but no longer stands. From the 1960s, two stone houses, one at the main pool and another at 'Chili',[4] were built and are associated with the families of the men that built them. Remains of much older dwellings with associated animal corrals are on the eastern edge of the *vega*, at 'Lila' where herders collected *pingo-pingo* (*ephedra chilensis)* for firewood, and at the drier pasture and salty springs of Tilocalar.

Tilopozo is lush and green, especially relative to the arid landscape on all sides. Herders like Inés told me that without animals grazing, the pasture becomes *suicio* (messy). There can be too much vegetation, the *pozo* becomes overgrown, making it dangerous; donkeys had been found in it drowned. It needs looking after. Perhaps responding to these concerns, in early 2017, Saturno, visiting the *pozo* with researchers, set fire to the vegetation as elder relatives had shown him. Recounting this, Inés described how the whole village, the fire brigade from San Pedro de Atacama (100 kilometres to the north), and the lithium mine's managers and workers got a fright.[5] The *vega* and everyone else had become unaccustomed to seasonal burning and the fire burned hot and high. Afterward, people went to have a look and, Inés said, the vega was '*bonita, limpiecita*' (nice and clean) and, what's more, beginning to send out fresh green shoots. Later that same year, they admired how 'clean' the *vega* and *pozo* were, relative to previous visits, in which thick vegetation obscured the ground and choked the channel leading away from the pool (see Figure 2.1 and 2.2).[6] According to Ines, the channel was a matter of communal work and responsibility, ensuring the vega would be nourished by spring flows.

Figure 2.1. Tilopozo, 2017 after burning. © Sally Babidge.

The production of the *vega* at Tilopozo by channelling and cleaning marks this practice as both management and ecological production, a form of practical relations between humans and non-humans that is common to herder communities of the southern Andes (Prieto 2015). Herders and their animals nurture (weeding, harvesting), feed (fertilization from droppings) and care for (do ritual for, accompany and watch over) the places made possible by springs, thereby expanding areas of pasture and seeking to ensure the existence of pasture into the future (Garcia 2021; Garcia, Prieto and Kalazich 2021). Tilopozo's ecological history may thus be told through the substantive effects of and connection to human and animal activity and practice as consequential for its shape, the quality of soil and pasture, and the groundwater that surfaces.

The water of the pozo is mineralized. In the 1950s, a pharmacist from the city of Calama wanted to bottle Tilopozo's water and to establish a health spa in the *vega* with tourist potential (see also Mostny 1954). While his idea never took off, the water's healing properties generated a community project started in the early 2000s, when, with visions of a post-mining future, they marked out a single road that could be used to reach both Tilocalar, with its archaeological sites, and the *vega* Tilopozo. Near the pool, they built a shade under which visitors might picnic. These

Figure 2.2. Tilopozo? c.1950 (image courtesy of Museo Nacional de Historia Natural, Chile).

first steps to create a locally managed tourist site also sought to address the problem of too many different vehicle entry points to the *vega*, with uncontrolled and damaging access to its surface, as well as imagined a future in which locally managed tourism would tackle economic concerns about life after mining.

In 2014, when I asked to film an interview with former community leaders Dona Valeria, Don Geraldo and Don Humberto, about the potential effects of mining industry water extraction, they suggested we do so at Tilopozo. As we sat in front of the camera, under the shade structure the community had built from *chañar* (*G. dicorticans*) timber and cane, I asked the kind of question one might perform for a recorded interview; 'How is Tilopozo important?' Some of what Doña Valeria responded spoke to Tilopozo's place in the future of the community.

> So, one day mining will finish and we will be left in a bad way if this place is dried out, if we need to go back to what we did before, looking after animals, herds of animals. We will bring them here to look after them, but we can't if the vega is dry. Well, they say that there is no damage, but we think there is. Because if they are extracting water from above, and they are extracting water from right here . . . [gesturing south] It doesn't rain much anymore, not much snow falls, so how will the water be maintained? It can't be. It's likely they are drying out the vega, because it would be difficult for this not to be the case with all the water they extract from above, from all around, and there is not enough

> rain and snow to maintain it. So that is why we have always fought with Minera Escondida. But now in the last meeting, they said they would begin using sea water and that only for a short period more would they continue to take water from up there [Monturaqui]. This is what they said in the most recent meeting. Well, is it because they have diminished the water here, or in good faith? I don't believe that very much.

The temporal, material and epistemological problems of Escondida's extractive relationship with the Salar de Atacama and Likanantay communities tumbles from Valeria's statements. With extensive industry freshwater use generally and Escondida's in particular, and a history of environmental harm evidenced by industry (particularly copper) in other locations, any economic life derived from the mining industry, or ecological life with the *vega* cannot be relied upon into the future. Too much groundwater, Valeria says, is being extracted compared to the amount of rain and snow falling. As she says, Escondida releases public reports on 'sustainability' and company employees meet with Peine community members to report on environmental 'impacts', but few people take company science as reported in 'good faith'. What happens in Tilopozo, as the principal area of surface freshwater and pasture close to Peine, matters in material ways.

Nor did the territory claim registered with CONADI by the Atacameño Community of Peine ensure Tilopozo's protection. In early 2017, some Peine community members responded to the news of Escondida's proposal to continue pumping groundwater by taking herds of animals to pasture in Tilopozo. I asked Inés to explain the decision to do this. She recounted,

> We were worried that eventually the *vegas* would dry out and there would be no pasture for animals. Because when I was a girl, there were many animals at that time. Mules, donkeys, and people walked with their herds. Many took their animals to the pasture there. So, I grew up seeing those things and taking the animals there with my mother and my father. When I was a girl, I went herding with the animals and I went with my own children later . . . So, my daughter said to me, '*Mami*, why don't we go now and take your animals to that pasture?' 'I don't know', I said, there are so few sheep now, but all the same one might want to relax, to stay a few days there, one could go. So, we decided to go. 'Yes', I said, and we were decided.
>
> There was a meeting in the community about this as well. About the vegas, about the water, everything. So there someone proposed, 'why don't some of us that have sheep go there and stay for a week, or if you want, more?' So, we invited people to do this. I was talking with another señora, with other señoras, and we said, 'let's go, let's go', maybe pass a few days there. And then they didn't want to go because they couldn't for all the things they had to do in

> Peine. And because I wanted to go, I asked my husband, hey, 'do you want to go?' 'Let's go', he said, 'let's go'. We told the people in charge of the community that we had decided to go. After that also another señor said he would go. And he went to Tilocalar. We went to Tilopozo.

Inés' and others' decision to respond to the Monturaqui EIA took her own and the community's fears for the place and its future to the source of their concern. Beginning more than four decades before, herding as socio-economic basis for survival had been interrupted by a combination of political, climatic and economic events. Now, taking the flock served a different purpose, performing the relation between community members, their herds, the pasture and spring. Inés' account gives voice to the competing demands on community members' time ('all the things they had to do'), a shadow of Inés' own worries about the competing responsibilities of contemporary life, and a reminder of the multiple ways in which people in Peine make life, with only a handful interested and prepared for a future livelihood as herders. Who knew if their trip to the *vega* would be effective as a protest about groundwater extraction? Community plans to go there, and just being there, seemed to be her point. This existential fact is perhaps best expressed in her account of the rhetorical question, 'Why not?', which she remembered asking her daughter, 'why should we not go and relax for a few days?' One might enjoy time spent at Tilopozo, since a person has rights in both the pasture and the water, and in doing so, keep the place company and perform the express desire to act on concerns, with their bodies and those of their animals.[7]

Scant surface evidence at Tilopozo of environmental harm from groundwater extraction, lack of independent studies to check company reporting regarding aquifer health and aquifer recharge, and the fact of vast and licit quantities of groundwater extracted each year, combine in community uncertainty, fear and tensions about how to respond to company claims of sustainable extraction. For those who took their herds to Tilopozo in 2017, they matched uncertainty with a statement of care for the place and claim over its future by taking themselves and their animals there, to enjoy it.

In what follows, I examine the problem of regulatory and scientific unknowns about water flows and volumes in the southern Salar de Atacama, and demonstrate how studies of ecology, vegetation cover, and hydrological connection between the *vega* and the underground are intimately associated with a kind of politics of the written material (Barry 2013). We know from other examples that extractive capital has manipulated truths in corporate reporting on environmental and social impacts of industrial activity and may actively obscure reports of serious harm (Benson and Kirsch 2009; Kirsch 2010). Ecological and social harms may be also sustained through complicit or effectively absent government regulators (Barandiarán 2018). The details and implications of a combination of material and political dynamics course through the remaining parts of this chapter.

The Capacity of an Aquifer

In the UN World Water Development Report (United Nations 2022), groundwater is said to have 'the capacity . . . to offer various services and benefits to societies' according to groundwater's 'geographically varying properties' and the impacts from human use and exploitation. As discussed (Chapter 1), the idea that water's active capacities may service human society is part of a universalizing capitalist animism that casts the non-human world in terms of its vitality as resource value. 'Ecosystem services' animates the non-human world as inhabited by so many different sets of resourceful activity, each with potential to feed economic calculations of productivity and potential for extractive growth. However, groundwater presents a problem in such rationality. It has long been imagined as having endless capacity, recharged in the infinity loop of the hydrological cycle. As technology has allowed, groundwater became increasingly drawn upon and while pumping technology has improved, groundwater science does not seem to have kept pace. Industrial (especially mining and industrial agriculture) capital has expanded its vertical axis of accumulation almost without impediment and now, as indicated by the UN report, groundwater is being exhausted. The aquifers in the Salar de Atacama are a case in point.

Hydrological descriptions of the Salar de Atacama begin with the fact that it is an 'endorheic basin', a geophysical subject considered 'closed' since waters of the basin do not flow on to rivers or the ocean. In the north of the Salar, the San Pedro and Vilama Rivers nourish the agricultural *allyus* of San Pedro de Atacama, perhaps replenishing the deep viscous salt lakes and scattered freshwater springs, when the river's flow is not siphoned off for mining (upstream), agriculture, domestic enjoyment and tourist hotel use. Average rainfall for the region is so low that for many years people throughout the Salar maintained *paja* or thatched straw and packed mud roofs. Violent storms annually since 2015, in February and March, have seen many of these covered with sheets of thick plastic or replaced with corrugated iron.[8] In the southern Salar, there are no rivers, and 'recharge' of aquifers is thought to begin with flows from snowmelt and rain from the Andean ranges, trickling down subterranean channels. Along their way, some waters rise as springs in the foothills. In Peine, from the lookout at the Cerro de la Cruz (a wooden cross on the hill), or when viewing satellite images, a dry delta in the soft sand (*playa*) can be seen fanning out at the bottom of the village; furrows left by flood and agricultural runoff water. These traces that water leaves mark the desert surface. Between Peine and Tilopozo, water rises to shallow brine lakes, La Punta and La Brava, with Salada and Saladita Lagoons to their east. Once in the salty surface and vegetation, water is subject to the clearest skies in the world and 'discharges' from the earth into the air, in the hydrologic cycle's evaporation (and from plants, evapotranspiration).

The description of the Salar as 'closed' elides extractive intervention in the water cycle. Government diagnostic reports, which included calculations of the average hydrological 'balance' for the years 1970–2008 (DGA 2010) also exclude calculations of extracted water. The Salar de Atacama 'basin' is a geomorphic and regulatory subject, with imputed boundaries between one area and the next subject to scientific study or political decision (cf. Orlove and Caton 2010). The DGA's 'balance' used rainfall data, available estimates of subterranean and surface flows, and evaporation and evapotranspiration data to present a calculation of how much water 'naturally' moves through the cycle (DGA 2010). As in other studies, the DGA found that the range of variables needed for a calculation of the flows in the Salar basin made their calculation subject to 'great indeterminacy' and 'uncertainty' (2010: 96). Nevertheless, the report provided a model, or an estimation tool for calculating the balance, and therefore a figure for water 'availability' (quantities of water apparently present in the basin at any one time) (2010: 126). The DGA recommended that its model might *in future* consider industrial extraction quantities, which largely occur in the southern areas of the Salar, identified in the report as the 'most sensitive'.

While lands around Tilopozo were registered as 'protected' by an act of the Chilean Water Authority (DGA, Resolution 425/2007), the groundwater available in its immediate surroundings and 'upstream' is the subject of extraction by Escondida and neighbouring copper mine Compañía Minera Zaldívar (owned by Antofagasta Minerals), which has fields of groundwater wells slightly downstream in the same aquifer as Escondida. In addition, freshwater pumping by one of two lithium extractors (Sociedad Chilena de Litio, later Rockwood, now operated by US lithium giant, Albermarle) occurs at 'Chili', adjacent to the main pool of Tilopozo, and brine pumping by Albermarle and SQM also occurs across the southwestern Salar (see Table 2.1).

Table 2.1 is a compilation of public data from the DGA's register showing concessions to water in the area as owned by mining

Table 2.1. Groundwater (freshwater only) extraction rights owned by copper and lithium mining companies, southern Salar de Atacama.

Company name as registered	No. Points	Volume (litres/second)	Years of concession
MINERA ESCONDIDA LIMITADA	25	1542	1987, 1994, 1998, 2000, 2001
MINERA SPENCE	3	450	2000
MINERA UTAH de CHILE	22	340	1985
SOCIEDAD CHILENA DEL LITIO LTDA.(SCL)	2	23.5	1983, 2000
C. MINERA ZALDIVAR SPA	17	625.3	2009
SQM	12	547	1991, 2012, 2014, 2016
TOTAL		3527.80	

Note: Compiled from public data, DGA website (https://dga.mop.gob.cl/productosyservicios/derechos_historicos/Paginas/default.aspx, last accessed 18 June 2023). Minera Utah de Chile was the exploration venture that founded the Escondida mine before BHP purchased the majority of the operation. Minera Spence is also operated by BHP. SQM's water extraction is the furthest from Tilopozo.

companies. The DGA grants rights to extract at the average rate indicated as a 'litres per second' (l/s), the number of extraction points (or wells), and the years in which different sets of wells (and their volumes) were registered. Rather than consider the data in table 2.1 as representing actual water quantities extracted from the southern Salar de Atacama, and thus a series of numbers that might be inserted into an equation such as that of the DGA's desired 'balance', the data is at best an indication of a portion of the legal potential of extraction activity. Some of the points may be inactive or outside the area; for example, in the case of the 340 l/s of Minera Utah de Chile, the DGA lists these as within the Salar de Atacama, but also as located further south on the border of the Salar de Punta Negra basin. Table 2.1 also omits the fact that there are a significant number of other extractors in the Salar de Atacama basin. Nonetheless, the distinction between a proportion of legal potential and actual volume or quantity is pertinent, since the register itself is uneven in terms of information available for each point of extraction and field of wells. These concessions are legal rights as registered, expressed in average annual flow volumes not actual extraction quantities measured by in-situ gauges. The DGA does not register saltwater or 'brine' extraction volumes as water, but as mineral, meaning that the many thousands of litres of water as brine extracted by lithium mining would be outside of any water availability model. Building a hydrological model for the Salar, it would seem, would necessitate combining a range of potential and actual ecological and extractive activities.

Few independent studies have examined the activity of groundwater for the whole of the Salar de Atacama or attempted to compile sets of data on the environmental harm relating to groundwater extraction for mining throughout the area (Babidge et al. 2019; Donoso et al. 2020). Hydrological research on the 'recharge and discharge' of different groundwater systems that feed springs and aquifers in and around the Salar, have referred to these as not well understood (Kampf et al. 2005; Messerli et al. 1997; Tejeda et al. 2003), and quantities of water that flow to salt and fresh lagoons on the Salar's surface have been described as largely unknown (Vasquez et al. 2013). Hydrologist John Houston concluded (twenty years ago, as corporate extraction concessions were growing in number and volume) that evaporation is a process that is 'elusive and poorly quantified', especially in the Salar de Atacama, and while he made a detailed attempt to address this, he concluded that 'important gaps' in scientific knowledge of the area remained (2006: 402).

With the projected growth of lithium mining in 2012, the DGA was scrambling to put together a 'diagnosis' of the Salar's environmental health. I attended a public consultation hosted the DGA and facilitated by a consultant group in San Pedro de Atacama, in which preliminary findings of the diagnostic process were presented to the public in terms of addressing the need to better regulate groundwater. Atacameño community leaders from villages and *allyus* around the Salar, as well as non-Indigenous residents, tourism business owners, journalists, and researchers such as myself attended the meeting. The DGA's report emphasized its limitations; a range of problems in representing and regulating the basin included the DGA's own network of environmental monitoring apparatuses, composed of instruments and data points. These included a proportion of data drawn from the instruments owned by mining companies. The meeting was told that 'the data, together, do not make a coherent set', that problems with the weather were responsible for missing data and discontinuous registers, and the presenter's concluding slide stated that the DGA has 'general problems with [data] reliability' (DGA 2012).

Some four years later, as part of a national investigation into glaciers and saltpans (Cámera de Diputados 2017), submissions to the report revealed the DGA had in fact diagnosed the Salar as over-allocated (*sobre otorgada),* meaning that according to the authority's calculations, there were more concessions granted to users of the water than water entering the basin. The submission quotes a DGA report from 2004, in which 'zone 4' of the Salar – the southern area where Tilopozo is located – had an imputed rate of natural recharge of 880 litres per second (l/s) and the DGA had granted 2,466 l/s to extractive capital. If we were to think for a moment *with* such modelling, this leaves some 1,586 litres extracted per second from only the southern part of the basin, of only freshwater (not including brine), which was not able to be replenished by imputed recharge flows. In even earlier DGA reports, similar figures appear. Statements by the water authority include that 'the effective flow of granted [extraction] rights exceed the renewable resource in this zone' (DGA 1999: 9). Despite such findings, the DGA public register reveals that in the year following that report, they granted concessions to the companies Escondida, Zaldívar and SCL in the southern 'sensitive' zone. A consultant who worked for the Peine community also made a submission to the national investigation (*Comisión Investigadora Glaciares y Salares*, see Cámera de Diputados 2017),[9] referring to his own study using satellite imagery to estimate the size of the wetlands and lagoons in the area of Tilopozo between the years 1979 and 2009, which showed that these had substantially diminished. He refers to a DGA report from the 1990s, in which the author noted the quantity of extraction by miners Escondida, SCL and Zaldívar, but claimed that since mining will finish in twenty-five years the subterranean flows will recharge the basin once extraction ceases.

According to a submission to this inquiry by the regional head of the DGA from Antofagasta, the government regulator was striving for better regulatory powers and seeking to build their technical expertise. His submission notes,

> The DGA can work with these aquifers and how they are used, only according to its economic capacities. Hence there are [water extraction] rights in this sector and they are increasing, but it [the basin] needs an 'SDT' [water availability model] to determine the state of the aquifer, how it is behaving, and how it is recharging. We must know the availability [of water] and for this we must have an SDT, which is not unique and will be variable. (Cámera de Diputados 2017: 96, see also 88)

The mixed messages of different actors from the government agency – that they have a model and estimate of 'recharge', yet they need a better model – can be understood as public acknowledgment of the problems of their estimations, the limitations of their own practices regarding regulation and allocation of water extraction rights, as well as the problem more broadly of the fact that groundwater had been allocated and extensively exploited by mining companies in this area long before it was effectively studied or regulated. Such matters are confounding against the DGA's activity to underwrite continued extractive interests. Compare, for example, the figures reported above with those of a 2014 report: that water rights granted through the Salar de Atacama combined to a total amount of 2,427 litres per second, yet an estimation of only 403 litres per second of natural inflow (DGA 2014: 74). Similar to his submission to the report cited above, in an interview in 2015, the regional director of the DGA explained to me that the Salar's remoteness from the DGA offices in the regional capital of Antofagasta (where we were speaking), its geophysical size, the water's location underground in an arid landscape, and problems with their environmental models all prevented the DGA from regulating water or knowing the true environmental impact of mining company extraction. But if extraction of water was occurring at the rate of rights granted and the modelling the DGA used to consider inflow was correct, why did the government agency allow further extraction concessions in the Salar, leading to over-exploitation by what some estimated to be around six times the 'natural inflow'?

The Director of the Water Authority had complained that the Salar is remote from his offices, too geographically extensive to properly monitor and that a complete monitoring operation was beyond the technical expertise of his agency. Each is an immediate structural and economic factor which belies the broader politico-legal conditions of Chile's private concessions to water, and the associated inability of government agencies to function as centres of knowledge and regulatory production against the private ownership of water and water data by mining company extractors (Barandiarán 2018). The DGA has had little oversight in the quantities of extraction and monitoring of associated impact. An 'absent presence' of the state (Bainton and Skryzpek 2021), in which regulatory unknowns of groundwater go unaddressed yet corporate extraction is actively enabled, is potentiated by the aquifer's material capacity for 'invisibility'. Groundwater's capacity for active invisibility has made it possible for Escondida and other extractive capital to continue to insist that groundwater is manageable ('sustainable') through assertions of industrial and technical competence. Like Escondida, each of the four principal extractive companies has their own mathematical model of the water, measuring the recharge, their extraction and the flow of water. Thus, each company presents their environmental models to explain how they manage the aquifers they draw upon and the potential harms of doing so. These models reveal further the specific ways in which groundwater may be unknown and apparently unconnected to the surface, while simultaneously voluminously present in registers of abstraction, as well as politically fragmented, making material calculations of its capacity difficult. It is to these corporate modelling processes that I now turn.

Complex Models and Company Monitoring

Despite company and state insistence on the lack of conclusive data, Peine community members have long insisted that Escondida's extractive activities are doing harm, whether they have done so in closed negotiations with company actors, such as those regarding an agreement for 'development benefits' or expressed as public protest, such as those mounted against new proposals such as 'Pampa Colorada', or in public media reportage about the damage Escondida wrought in the Salar de Punta Negra (TVN report, 2007, as discussed in Chapter 4). In engagements between Escondida and Peine that I witnessed, community members asked, 'when will Escondida tell the truth about water extraction damaging the vega?' Company employees responded at different times and in varying ways to community concerns. Nonetheless, up to the 2017 Monturaqui EIA, employees from the company, and their reports, consistently asserted that there was no permanent harm being done to Tilopozo as a result of their extractive activities.

A key element of Escondida's environmental reporting activity was a model the company scientists devised called an Early Warning Plan (*Plan de Alerta Temprana*) or 'PAT' (Minera Escondida Limitada 2010).[10] According to Escondida's PAT documents, the impact zone of Escondida's water extraction in the southern Salar overlaps with Peine's recognized territory at Tilopozo, which is also a DGA-protected area. The PAT, the company created for the sector it calls Monturaqui-Negrillar-Tilopozo (MNT), was approved by the DGA in 2001 and contains the model by which Escondida publicly represents the processes and effects of company extraction. The notion of 'early warning' relates to a planned program of collecting monitoring data, which, as it is inserted in the model, would 'alert' the company to impact on a protected environment if certain indicators were reached. The company scientists created these indicators according to the data they had collected and evaluations of the MNT environment. Thus, the company and their neighbour Zaldívar have rights to a total of over 2,000 litres per second in the sector (table 1), though these volumes appear in the PAT as a maximum rate of water extracted from the sector at 1,800 litres per second. The second indicator of 'early warning' is a decrease in the level of superficial water throughout the sector to a maximum of 25 centimetres, and the third indicator is a 6 percent maximum drop in the flow of water underground from Monturaqui-Negrillar to Tilopozo.

Between 2010 and 2012, and to comply with their Company-Community agreement, Escondida employees used the PAT to present environmental impact data to the Peine Community at meetings in the village *sede*. For these three consecutive years, Escondida's social performance manager made a special effort, as he told me in a conversation in 2012, to assure that the reports were 'consultative' and 'transparent'. In the two annual meetings I observed (2011 and 2012), company speakers focused their reports on calculations showing adherence to the limits of acceptable environmental change in the MNT zone, as established in the PAT. Despite such assurances, from early 2011, the company sought to improve its modelling of groundwater activity and industrial impact through renewed campaigns of data collection. Company employees and contractors installed themselves in the southern Salar drilling bores. Escondida staff explained their interest in 'creating a [better] model of the watershed', in the interests of regulating extraction according to the limits set in their PAT.

In 2011, thirty or so men and two women from a drilling company contracted to Escondida were staying in Peine for the monitoring campaign. The *Ckunsa Turi* Guesthouse had been expanded fourfold with a bank of converted shipping containers.[11] The drilling contractors were taking months longer than expected, and Peineños began to air suspicions about what the company was doing. Over the months that I spent in Peine that year, unanswered questions circulated and led to heated discussion in a community meeting with company social relations manager and employees. People wanted answers. Why was the drilling company taking considerably more time than first planned and reported? How many monitoring wells did the company plan to perforate? Originally seven were proposed, but now it seemed like more. Were these wells for monitoring, as initially proposed, or was the company intending to extract water from these new holes as well? Why were there lights at night in the region of these wells? What were they doing there at night? Although the stated purpose of company drilling was for 'monitoring and measuring the watershed', Peineños suspected that mining company operations also involved new extraction. There was also some annoyance at the ongoing presence of this number of company people for so many months. Company representatives agreed that a group from the community leadership could take a day trip with the company scientist to the MNT zone, and observe where the drilling was occurring.

Escondida technical staff led the tour, showing the group cleared areas of land referred to as 'platforms'. Some of these were ready for drilling to begin, others held metal caps where drilling had already occurred. We watched a truck perforating a new hole, looked at old and existing extraction infrastructure and drove on gravel roads alongside large diameter black pipes stretching toward the horizon. The junior hydrogeologist from Escondida seemed to be doing her best to explain the industrial and scientific rationale for and the process of drilling holes in order to measure and monitor groundwater, but there was little she could say about closed metal structures, cleared dusty land and mute pipes that lay before us.[12] We approached Tilopozo from the southern side, making our way slowly to the *vega* across progressively muddy tracks until the Escondida vehicle became stuck and we had to winch it forward on the road using my jeep.[13] Eventually, sitting beneath the bough shelter, community members spoke about coming to the place when they were young, with donkeys and sheep, and their plans to upgrade the road so that the location might be used for tourism and recreation. The older people said they needed to have greater control over the area, to stop so many vehicles entering the *vega*, driving in all directions, causing harm to the ground, plants and animals that lived there. Don Bartolo and I stood at the *pozo* and peered into the blue green pool. He noted the weeds and algae, but despite all this talk of impact and monitoring, he said, Tilopozo 'still looked the same. But how could we know?'

In the following year, in the second annual company-community meeting I attended in Peine, Escondida scientists reported on the model they had created, and which needed consistent data input to monitor groundwater, given their rates of extraction. Company employees proposed that in addition to the monitoring 'wells', they had sunk throughout the southern Salar and the 'MNT sub-basin' during the previous year, they would monitor the world of the aquifer beneath Tilopozo and the salt lakes using 'telemetry'. This involved installing monitoring instruments, which would send electronic data to Escondida scientists. Telemetry, the company scientist explained, would be a better form of monitoring the water levels underground and in the lakes and pools of the southern Salar. It would also be a method with much less impact than other forms of monitoring; less taking samples for monitoring meant less driving around in the environment and less bothering the community for permission every time they wanted to do so.

At the time of the proposal, I asked Doña Rosa, a former community president, how she understood what was proposed, and she explained to me that,

> They want to do other kinds of monitoring, so that they can read their wells right from where they are [in their offices in Antofagasta]. These [new instruments] are on tall poles, with a panel on top that would sparkle in the sun. So we said no. Now they want to put them on the ground, and we are in conversation about this, about what they are really trying to do and what we will respond. They say it is more environmental damage to continue to enter the area in *camionettas*, and better to be able to measure the wells from a distance. I guess the data goes to them by internet . . . ?
>
> [sb] So the Community doesn't want the telemetry instruments?

> We don't, no. Tourists might come and see these shiny panels and the animals and all that, because you can see them shining from far away, they are really tall. Now, we made them understand this concern and they say they can put them on the ground and that's as far as the conversation has gone now.

Rosa was patient in her explanations and conveyed her years of experience reading company reports and evaluating proposals against opinions expressed in her community, as well as in community-company meetings. Rosa argued for the need to have proper technical support themselves, people who could be *their* scientists, since she asserted, it is the '*tarea de la comunidad*' (community's homework) to monitor the environmental impact on Tilopozo. She explained,

> Well, we want someone who is an expert to come and measure the wells with them [the company] to verify if it is the truth or not. This is what we want to do, this is the kind of work we want done in the environment. But you have to contract an expert person who knows how to measure wells and see if what they say is okay or not. If we do this, we will know exactly if there have been drops in water level or it is being maintained like they say. Sure, Escondida exhibit their reports [*scribbles in the air with her hand*] but we don't know, we're not experts, we don't understand anything about their lines that go up, go down, are the same, nothing.

That is, the only solution to the fact that they don't trust company scientists and company data is to engage their own experts to examine the way data is collected by the company and to check up on mining companies when they do their studies.

Without scientific training, like Rosa, and unlike Rosa, without local knowledge of Tilopozo, I found it difficult to effectively evaluate company reporting in the meetings I observed, or during a day out with the monitoring team, looking at drill holes in the landscape. In the meeting I attended the following year, I asked a naive question at the end of the company scientist's presentation. He had displayed PowerPoint slides of graphs and other forms of data visualization to provide his assessment that groundwater levels in Tilopozo were 'stable'. I believe that this was the first time I had ever spoken in an *asamblea* apart from explaining my own research proposals to the community in the previous year and, excusing myself, asked if he could explain a little more about the PAT model and the nature of stability. He responded that the modelling is complex process and couldn't be explained in the meeting. Nicolás, manager of social relations at that time, tried to soften his colleague's response by stating that their model was approved by the DGA, but that their aim was to constantly improve their scientific models.

In 2015, I included in my planned fieldwork a trip to Antofagasta to speak with company employees and scientists. After a short conversation with the then new manager of social relations, I was granted an interview by a hydrogeologist at Escondida, who described the scientific problem of modelling groundwater, as well as the challenge of explaining their modelling work to the public, especially those concerned about the company's environmental impact. She demonstrated her concern for the problem of scientific communication in her careful description to me of the Salar's hydrogeology:

> The Salar, any Salar, is a system of discharge and the water discharges through evaporation. It [the groundwater] may have a connection with another sector, but by preference, *el agua se va a querer ir por allí* [the water is going to want to go that way]. Because water, I always say that water is super-intelligent, it will always want to go for the easiest way and the easiest way is to discharge through the Salar . . . If here we have some clay, for example, *va a llegar aquí y decir, no puedo pasar, tengo que irme por acá* [the water will arrive there and say, I cannot pass, I'll have to go this other way]. So, for us it is really important to know the geological formations. Because we know – well, we infer, obviously, we can't say with scientific certainty – *inferimos* [we infer] . . . which way the water is going to go, and in this way we build our model. We try to explain this also to communities sometimes, but sometimes they do not believe us very much [laughs]. *Pero así es* [But that's just how things are].

The scientist's personification and simplification of scientific modelling for the benefit of an interviewing anthropologist and an imagined community audience describes groundwater's activities and capacities much like UNESCO's lively account of what groundwater might do. The forms of uncertainty expressed are simply part of the scientific conditions of groundwater. However, these rationalities of scientific knowledge and its limits were not the mode of science reflected in company assertions, neither in their public reports nor in reports directly to Peine community members. In contrast, Escondida's MNT PAT is a model in which Escondida predicts 'sustainable extraction' with certainty.

Escondida's PAT modelled impact through imputed basin recharge, subterranean flows that begin in Andean peaks and measured at various monitoring points, and discharge, evaporation and evapotranspiration, measured in Tilopozo and the nearby salt lagoons, and volumes of extraction by Escondida and Zaldívar. While Escondida's hydrogeologist explains that the first and second aspects of the model are inferred, in fact all three calculations are necessarily arbitrary. These matters were not lost on the government regulator. A DGA report in part notes that,

> Of the three indicative conditions in the PAT . . . it is possible to indicate that the first of these is considered ambiguous, given that the 25 cm level has no established reference point . . . and does not indicate which monitoring wells were used to verify [change] . . . or the period of time that determines the threshold. (DGA 2014: 250)

Each aspect of the PAT relied on studies produced almost solely by Escondida's scientists, rationalized, one might assume according to the groundwater concessions already and anyway granted and used. According to the DGA report, their analysis of Escondida's monitoring wells indicate that the levels of the PAT were 'closer to the established threshold' than the company reported (DGA 2014: 217), thus indicating inconsistencies in company data, and contrary to company claims of stability, that a steady decrease in water levels at Tilopozo was occurring. Moreover, the DGA report estimated that one of the early warning signs, a 6 percent reduction in the flow to the aquifer as identified in the PAT, may be reached in only four to five years (2014: 251–55). When I spoke to the company scientist that I quote above, she told me that if a significant impact on Tilopozo is registered, it will occur a 'long time in the future'. This is a rather different conclusion to the four to five years the DGA predicted in its report, based on company data, the year earlier. As far as I could find, the company's PAT does not provide ecological reasoning for its three 'alerts', and no reference is provided for how the company measured recharge or other ecological factors outlined by the hydrogeologist. Despite all of these matters, the government agency could not prevent Escondida and others from continuing to extract up to the end of its water concession in 2019, and at the levels it had imposed on itself as sustainable.

The conflicting representations of the 'MNT' between Escondida and the DGA describe how the relation between extracted groundwater and Tilopozo's water and ecology has at times appeared well established, however, the details of the connection are subject to many unknowns. Escondida's continual attempts to gather data show that knowing the connection, or appearing to have good knowledge of the relation is necessary for their claims of sustainable extraction. Using their own data and eliding that produced by the DGA or independent scientists, the company continued to assert the 'sustainability' of its extractive activity and the community continued to doubt their assertions. The PAT model defines and encapsulates a common corporate risk strategy, which, as McGoey argues, represents 'commodified uncertainty . . . a calculation of one's exposure to contingency . . . itself contingent on its own inability to predict all contingencies' (2012: 8). Escondida's assertions of the stability of Tilopozo and negligible impact from continued extraction was a risk the company took in trying to ensure their continued access to waters of the 'MNT' area. The effect of this commodified uncertainty has been a manufactured ignorance about the capacity for groundwater to be effectively related with future ecologies in the southern Salar. The company's official assertions about certainty and sustainability were, as it has turned out, a harmful promulgation of unknowing about the ecological effects of extraction, both in reports to the Community and in those to the government authority.

Simple Facts

Scientific activity has followed the emerging lithium boom since about 2014, somewhat responding also to the lacunae in knowledge of the Salar's ecologies. Hydrologist Marazuela (with colleagues) published a series of articles that consider the hydrodynamics of the salt flat, modelling freshwater and brine activity and proposing 'sustainable' extraction models based on their data (most recently, Marazuela et al. 2020, 2021). Seeking to address the epistemic gaps, especially given the threat from brine extraction, biologists report that the Salar de Atacama is 'dominated by microorganisms', hundreds of species of 'extremophiles' that live in the Salar and survive in a 'lithium soup', a hypersaline environment, that may help humanity better understand life in extreme environments (such as Mars) (Cubillos et al. 2018, see also Thiel et al. 2010). Chilean scientists also pronounced a 'microdisaster' of the Salar de Atacama given the impacts to these microbial ecologies, connecting the impact on microorganisms to the long durée of human history (Bonelli and Dorador 2021). At another scale, conservation scientists have reported that water and brine extraction from the ecosystem of the Salar de Atacama has affected the brine shrimp (*Artemia*) and iconic species of Andean flamingos (Gajardo and Redón 2019). Others have begun to report a broad-based reduction in flora and faunal richness across the environment and the lithium industry threatens to pollute water sources with toxic by-products (waste) (Kaunda 2020).

As extractive pressures have escalated, the Peine community has employed a team of professionals and environmental scientists, as well as sought to engage independent consultants and scientists to undertake studies of the southern Salar. In 2017, one group of scientists from a private research institute was interested in undertaking broad-ranging environmental monitoring and training Peine community members to work alongside them. However, this project, like many others, did not eventuate in the way envisaged by either. A problem of monitoring the company, or monitoring the environment, can present disjunctures between what the community may see and discern in places they are related to, like

Tilopozo, and how they translate these relationships into their practice with the surface and subsurface of reported environmental objects subject to extractive activity, making their knowledge recognizable to companies and regulators.

In broad terms, Escondida claimed it was managing the impact specifically on Tilopozo and also on Indigenous territory, but despite the efforts of different teams of company employees, these claims were not trusted by members of the Peine Community. Partly this is reputational, since the harm caused by Escondida's extraction in other places informs how people talk about the ongoing risk for Tilopozo. In discussing such matters further with Don Geraldo (who, as he asserted, has been involved in meetings with Escondida 'from the beginning', by which I understand, since the late 1990s), I told him how I was having trouble understanding Escondida's environmental reporting and modelling. I said, the aquifer seems to be rather mysterious, making the effective relation between Tilopozo and Monturaqui difficult to prove. He rejected my suggestion of groundwater's enigmatic character, telling me the following story to illustrate his point:

> When I was young, we went to the cordillera with our animals, to that part called 'Leoncito' . . . The mountains in that area are very high and the snow is a long, long way away. Far below is a pool of water. But when there is no snow, the water dries out! [laughs] So there you have a test! I did a study! . . . It looked just like a spring had appeared. But when the snow melted, little by little the water dried out. *Listo*! [Right!]. Where did the water go? It wasn't running upwards; it went below . . . So, it makes one think about the currents of water. The fact is, they run down and they come from above [laughs]. Sure, it's just about seeing the way things are.

Geraldo's knowledge of the mountains, the Salar, and its waters is based on his lifetime of observations and knowledge of material connections. These are dynamic over time, yet have particular and undeniable capacities (to run down, for example). The basic principle of downward flow Don Geraldo noted in relation to Tilopozo was the same as the cyclical appearance and disappearance of water at 'Leoncito', with the example aimed at demystifying what cannot always be seen. It does not resolve many forms of uncertainty but dismisses technological arguments about what water might do if managed by extractors and denies that any other factor may be at play, except the obvious association between extraction and harm. It is simple; if they extract enormous quantities of water from above, they will dry out what is below. Embedded in this logic is the relation between one action and its effects as a matter of scale, as Doña Valeria said, *how much* water is extracted, and also the extent of historical reduction in water flows; what has been done to groundwater, and thus to Tilopozo is harm and depletion. How can that possibly be mitigated?

Such facts informed the ways in which the Community has always rejected Escondida's models of harm management. It might also be said to have informed how the *asamblea* in Peine's hall in January 2020, responded to the company's Monturaqui EIA. At the beginning of the chapter, we left proceedings at the time of the vote. The result was that the overwhelming majority voted 'no' to Escondida's continued water extraction, and 'no' to further negotiations with Escondida over the compensation measures proposed by the company. Following news of the vote, Escondida released a press report saying that it had decided to retract its EIA. The Antofagasta edition of the conservative national broadsheet, *El Mercurio,* reported that the company would abandon continued water extraction from all subterranean wells. How had the Peine Community managed to say no to a global giant and have their decision heard? In the chapters that follow, I delve backwards and forwards in time from this event to explore the answer.

Conclusion

While extractive capital has produced data and models and engaged in environmental monitoring with public vigilance, none of these activities have addressed ecological and political uncertainties of communities like Peine. This is partly because of the particular conditions of advanced extractivism in Chile, and the fact that capital obfuscation enabled by material politics has been largely unchecked and unregulated. Groundwater's material unknowns have been active in the material politics of uncertainty. Generational change among Community leaders have seen their arguments reach into regulators' decisions in ways that are surprising given the short history of mining relations. This has given Community leaders more support than before to push back on global giants.

Water's surface appearances afford flourishing; '*lo que aflora*', what surfaces, has an active relation to the good life in the past, told in ecological histories, and is related to people's agency in imagining and generating a good future. Water's 'relations' need not have a positive connotation (Strathern 2020). Since 2017, people have insisted on being able to see environmental harm, including that in Tilopozo, '*se aflora la sal*', what comes to the surface, is a glistening salt crust visible in the mud and pasture areas that was not there before. Relations may be conflictual, may sparkle in the sunlight and may thus be based in depletion or absence. Groundwater is connected to the surface, but the connection is interrupted by both extraction and the models of sustainability, formulated and executed with partial knowledge. The problem of groundwater is not then to make it visible, but to make more visible the threats to its exhaustion. The question that arises is, if water is being extracted by mining companies at so many thousands of litres per second, how can this water possibly continue to support the ecological relations that produce flourishing?

Notes

1. In a critique that reveals the broader socioeconomic failing of the Chilean neoliberal state, Barandiarán (2018) details how the SEA were not a well-staffed scientific outfit evaluating the merits of projects in the national social interest, but like other Chilean state agencies, acted only as 'umpire' between proponents and respondents. This is a term Barandiarán applies to the analysis of state forms in relation to the regulation of capital, and it is contrasted to an 'empire' state, in which agencies are built to develop scientific expertise in the national social interest. The critique is directly relevant to broader commentary on incremental reduction or outsourcing in neoliberal states of technical expertise.
2. I was also present. At the beginning of the process (2017), Peine's Directiva at that time requested that I undertake preliminary analysis on the sections of the EIA report categorized as 'human impacts', and to that end I undertook field research and desktop analysis and accompanied leaders of the Peine Community to the Santiago home of this lawyer, who agreed to provide the community with strategic legal advice on this matter. Like all such work I had done before and since, this was undertaken without financial payment, though in this case, my travel costs return from Santiago were paid for by the community.
3. Inference drawn from consultation of the *Field Guide to Aquatic Plants in Chile* (Rodriguez and Fica 2020).
4. Chili is the vent from which Rockwood Lithium/Albermarle extracts freshwater.
5. The fire threatened Rockwood's water extraction infrastructure (run on diesel pumps), others clicked their tongues about the *animalitos*, especially *chululos* (*Ctentomys fulvus*), that might suffer from the blaze.
6. The older image from the MNHN, I believe, depicts Tilopozo. This and another, of the village of Peine probably in the early to mid 1950s, were mounted on board and hung on the wall in the MNHN for many years (Pers. Comm. Director MNHN, Sr. Guillermo Castillo, 13 November 2017). It is assumed they were taken by former Director Grete Mostny or by the geologist E. de Bruyne. Dr Pedro Mege, gifted me his publication '*Andinos: Fotografías siglos XIX y XX*' which alerted me to the existence of these rare images of Peine and possibly its surroundings.
7. The verb 'enjoy' is reminiscent of the western legal phrasing to 'enjoy rights in –' .
8. In San Pedro de Atacama, those interested in preserving the character of the town have devised a method of sealing with plastic or tin underneath the adobe exterior.
9. The consultant is interviewed by the Senators of the Inquiry for around 45 minutes (see https://www.youtube.com/watch?v=R4zDluBL3uM, from 6:15min, last accessed 19 June 2023).
10. Worth noting that company scientists and social relations managers I spoke to emphasized this; that Escondida invented the PAT instrument.
11. See Chapter 4. In the 1990s, the Community built the Guesthouse in an effort to attract some of the tourism traffic in the region.

12. One older leader complained that sending a junior staff member to respond to community members on this fieldtrip was a typical example of how the company misunderstood the need for adequate communication.
13. I participated in the day trip and was given lunch to take along with me like the other participants. I drove my own vehicle, paid for the fuel and carried two additional passengers from the community. I also wrote a brief report to the Community describing and summarizing the day's activity, what we encountered and photographs and GPS readings of the drill holes.

Chapter 3

AGUA DULCE

The Labour and Logic of Good Water

Two principal springs sustain Peine's agricultural oasis. These are registered in the Water Authority (DGA) as Vilti and Puri, and they produce waters with a strong mineral taste that run a little warm. These flow into the uppermost two of three open reservoirs for agricultural irrigation and an inviting pool for recreational bathing. There is nothing wrong with Vilti and Puri's waters, people told me. They have 'always' fed the oasis vegetation, crops, animals and humans, but beans don't cook well in them, powdered milk doesn't mix properly, and the waters are 'salty'. Given these material properties, sometime in the late mid-twentieth century, people began to talk about finding a new source of *agua dulce* (freshwater). The epic that unfolds from the decision recounts impressive deeds achieved through communal effort. It was narrated to me by numerous people, especially when I asked about social change associated with their experience of the beginning of mining industry. Almost everyone began their account of these matters with the fact of a collective decision that they must address the insufficient quality of Vilti and Puri's water, as evidenced in tough beans and split milk.

This chapter traces some threads of Peine's story about harnessing the spring called Chakizoke to provide a new source of domestic freshwater for the village. As the community appealed to the state for support to improve their water supply, the story becomes entwined with the temporalities of national politics; as an incipient project during Allende's short term in office (1970–73), put on hold with the dissolution of many forms of socioeconomic possibility during the Pinochet military dictatorship (1974–1990), and then revived with Chile's return to democracy in the early 1990s. Finally, securing Chakizoke through their own hard work, a small but insufficient grant from the state, and a range of assistance from

Notes for this chapter begin on page 107.

mining companies, the story then seems almost a parable of the empowering possibilities of self-reliance under neoliberal governance.

That is the tenor of the documentary film Minera Escondida produced in 1997, titled 'Peine, the Efforts of its Inhabitants', which depicts how Peine community members captured the freshwater for their village, and that the company funded some materials and machinery. The secretary of Peine's community leadership and administration (at that time, the Junta de Vecinos, JV) appears in the final frames, orating how community members laboured together to produce a good future:

> [Sec. JV]: The people have been enthusiastic to work, to sacrifice, no? This sacrifice is important, no? It is good for us. More than anything it is good for our family, for the future, don't you think? Because one, unexpectedly, might need to struggle to work for the family. Because, we say, we are doing this for the family, for our children who come after. We are bringing this vital element at great sacrifice, we all know about that . . . What the people of Peine have is that they don't need people to come and do things for them. All we need is support! And *we* will do it. In this we are very clear, the people of Peine, *cuando se las pone la pila la pone*[1] [once they get going, they really know how to work] . . . and they do good things like we're doing them here.
>
> [Narrator:] . . . this [work] is setting an example for the rest of the communities of the Salar. (Minera Escondida/BHP c1997: 4:00–5:00 mins)

As the JV secretary asserts, the community's collective labour and sacrifice meant they achieved their goal of providing their families with good drinking water, with just a bit of support. The local narrative tells of community solidarity at a difficult time. Atomizing pressures of monetary wealth and temporal and social demands from wage labour in the nearby lithium mines affected some aspects of the mutual reliance and moral impulse sustaining collective labour practices. Aspects of these matters bring to mind Trawick's (2001) analysis of the moral economy of water in 'the Andes'; the ethical dimensions of fostering cooperation and solidarity through irrigation and water management. In the case of Peine, the freshwater project also further bound the community in socioeconomic relations with the mining industry and the state. In doing so the story of the freshwater project traces the emergence, remote from the state and centres of power, of the globalizing effects of extractive governance and its others.

Finally, the history of capturing Chakizoke as a new source of freshwater tells of struggles to install, secure and manage the infrastructure and the water it carries from spring in the mountains into reservoirs and through pipes stretching down the foothills eventually to the village grid and household taps (faucets). This history pushes into the present, with local struggles to keep hold of the water as source, material and flow; calculating its destinations, its quality and contents, and retaining its circulation to fortify local relations in substance and moral imaginary. The story articulates a moral judgement about the sense that, to be 'good', projects about the future are locally emplaced and controlled (cf. Blaser et al. 2004; Escobar 2012). The social and infrastructural events and practices I recount reveal people's judgements about what makes good water, and how this embeds their evaluations about what might make a good life.

Water as Infrastructure and Being

'*Se cortó el agua!*' (the water is cut) was a familiar refrain during fieldwork visits, especially in 2011. The phrase might initiate conversations in the street or in corner shops like a comment about the weather. Securing freshwater flows from Chakizoke into village homes had been riven with problems, the infrastructure struggling to be adequate to the landscape, climate, demands of government officers, or ever-changing needs of the community. During July that year, contractors to the Department of Water Works (DOH, *Departamento de Obras Hídricas*) were attempting to re-install, update and stabilize a water grid that, according to local commentators, had 'never' been adequately installed when the spring named Chakizoke was first captured. The contractors were working to excavate the rocky ground of Peine's streets. While they dug trenches, seeming to me designed solely to trip the unwary, and the grid remained connected to the houses, winter snow froze the pipes high in the cordillera. With the water cut, contractors were paid to truck-in drinking and cooking water twice a week. Residents secured 'IBC' tanks to dwelling roofs and onto the back of their *camionettas* ready for filling.[2]

The water from Puri and Vilti is good for making one's skin soft and animals prefer it over tap water. It is the right water for soaking and peeling corn for the festive dish *pataska* and belongs to a range of modalities of the present that sit well with a sense of the past. Women in their fifties told me about how as children they would collect water for their households from the *nascimiento* (lit. birth, beginning, appearance, in this a spring's appearance at the surface) or further down, from the cut-out pools along the irrigation ditches. As Doña Alma said, when she was a girl,

> We just had the water from here, ... *ese agüita salobre* [the little bit of salty water], that was our water. We didn't worry about water. ... We had to collect water from the little pool here. A heap of kids there, collecting, filling up jars, waiting to fill up their containers. And going along playing. It was a lot of work. *Ay, era más tranquila la vida* [life was more calm then].

A few communal faucets were installed in the village in the 1970s, and not long after the water was piped, as Alma said, 'right into the kitchen'.

In the early 1980s, village leaders also organised to purchase drinking water by the truckload from neighbouring places around the Salar, including Toconao, Talabre and Socaire, villages whose spring water is less mineralized. Don Tomás from the newly opened lithium mine (operating as SCL), who also needed to supply fresh drinking water for the worker's camp, helped by installing a standing tank, trucking the water in, and paying a share for the infrastructure and its contents.[3] Village households were allotted a time for filling a drum or two for drinking and cooking. Doña Alma said,

> A truck came from the company, from *El Litio* [SCL], it poured the water in there and we would go every week. My aunt and my son would take turns to go and fetch water. Fresh water . . . And that had to last a week. But just for the kitchen. We already had *agua salada* [salt water] in our houses.

In the 1990s, the municipality funded construction of bathrooms with flushing toilets built next to each dwelling throughout the region. As Alma explained, in Peine this infrastructure project simply extended the spring's 'salty' water directly into the home and people continued to collect barrels of potable water by the truckload. For some time, the dual sources of water provided basis for the desire for better water, as similar conditions did in other towns and villages of the Salar.[4]

Don Gerardo explained that the idea to improve their water came from a long-held frustration about water quality;

> It all started because we had just this water from here [Vilti, Puri]. We grew up with this water, but it is very high PH. Grains and beans don't boil, it cuts the [powdered] milk. So we thought, about '71, more or less, we began to be bothered by the idea of bringing better water.

While Don Geraldo described the water's high PH, others expressed their dissatisfaction in terms of the interaction of water with these alimentary elements – grains, beans and powdered milk - but also the drag of having to cart water. As Doña Alma said 'before we had to go and fetch water at the freshwater tank, . . . we didn't have potable water. No, that was a big change, uuff! Happiness, a great happiness!' Alma's joy, having freshwater right into her house, was echoed by many others.[5]

However, in winter 2011, the water was cut for months. In the house where I rented a room, cooking and drinking water was kept in the hallway in a tightly sealed barrel and carefully ladled out. Don Alfonso, president of Peine's Rural Water Committee (CAR, *Comité de Agua Rural*)

Figure 3.1. 'Arriba' showing the back of the shelter in late July 2011, with Miñiques peak in the background. © Sally Babidge.

scheduled a day of village communal labour to affect a work around to the water problem.[6] Each household was informed to send a representative to the plaza at 8 a.m. with any relevant tools in their possession. As the last stragglers of the 40 or so men and women hustled into view, Don Alfonso announced that two projects would define the day's work. The first would redirect the town's water infrastructure, disconnecting the grid from the lines that usually supplied fresh water from Chakizoke, and reconnecting to an old holding tank to which the water from Vilti and Puri could be pumped. The 'old' 'salty' spring water would suffice for the interim. This work, Don Alfonso explained, was both heavy and technical. Strong men were needed to drag the big pipes from one place to another and reconnect them, and a few women could help clear the ground of any debris. The second project would aim to finish construction of a shelter located on a plateau a few kilometres up into the foothills, a key site in the commemoration of the successful freshwater project with a clear view of the Miñiques peak, which feeds Chakizoke's waters (see Figures 3.1 and 3.2). The shelter with its stone seating area needed finishing in time for the first day of August and the '*Pago a la Pachamama*', a *convido* commemorated across the Andean region in different ways.

Figure 3.2. 'Pago a la Pachamama', 1 August 2011, Salar de Atacama in background. © Sally Babidge.

Don Alfonso called the roll of registered heads of households, and each person present responded by indicating whether they would stay in the village or go '*arriba*' (above). Jokes were traded among men and women as they answered. When Don Alfonso asked me what I was doing there, I answered as I'd been instructed (by the head of the household from whom I was renting a room): '*voluntario*' (volunteer). Juliana, with whom I'd been chatting, told Don Alfonso what to do with me by saying, 'as an anthropologist, she should go *arriba* . . . it is more cultural, better for her to learn'.

Thus, when the pipes with Chakizoke's water were frozen, it was not just infrastructure work that needed attending to. Chakizoke is 45 kilometres from Peine, adjacent to Lakes Miscanti and Miñiques, at about 4,000 meters above sea level. A portion of Chakizoke's water flow belongs to Peine in the legal sense, along with the litre per second quantity from Puri and Vilti registered as communal property of the Indigenous Community of Peine with the DGA. Each of these waters are held in Indigenous communal title, with Puri and Vilti 'always' having belonged to Peine, going back to pre-Hispanic irrigation systems, agricultural plots and the needs of the village residents. Chakizoke, within the territory of their neighbour Socaire, was subject to Peine community infrastructural intervention and registration of some of its flow since the mid-1990s.[7] Each of these spring flows are resources that people work for and with, and they belong in and partly constitute Atacameño-Likanantay territory. Each spring is also related to the lively being of particular peaks and related to people through the essential fact of the waters as associated with peoples' bodies, labour and lives. In speech, action, history, and ceremony, Peineños articulate their relations with Vilti, Puri and Chakizoke in terms of such multivalent relations.

To address the problem of frozen pipes, communal labour re-routed and managed infrastructure enacting community responsibility for vital resource provision. The project to move the pipes around in the village was coupled with an eye to ritual practice and both activities aimed to ensure water flows. At the shade structure, we wove lengths of wire through cane to make a neat roof, which was affixed to poles of *chañar* timber dug into the ground, and men sourced and hauled large stones to complete a bank of seating. Physical labour at the shelter made it ready for the ritual to be held a few days later. This water labour may also be observed as performing relations among community members in Peine, as well as to non-human beings such as mountains and waters, with relations to the state and mining companies partially present. What constitutes good water for locals may be parallel to and at times incommensurate with technical definitions of water quality, which in turn inform the definition of water as potable. As I discuss below, the definition of good water matters and it changes over time. I turn next to focus on the project of bringing freshwater from Chakizoke, which began before Peine became a legal entity, an 'Indigenous Community', and how the project laid claim to a local sense of modernity.

Knocking on Doors

Don Gerardo remembers that people in Peine began to be *molestados* ['bothered'] by the question of a new fresh water source in the early 1970s, a time of broader upheaval in ecological and economic life. Exploration for minerals on the southern Atacama saltpan had begun with the Production Development Corporation (CORFO, Corporación de Fomento de Producción) undertaking the initial studies for lithium and potassium extraction and had employed a handful of men from Peine whose knowledge of the saltpan's activity and contours was crucial to the surveying and building of roads. Rainfall decreased across the region and in lower altitudes vegetation radically diminished. Herders who had always moved between lush pasture areas in the broader territories were increasingly prevented from doing so by drought that made the paths between pasture and waters too risky and unprofitable. Greater numbers of young people moved away from Peine to find paid work. In 1970, Salvador Allende became the world's first democratically elected socialist President.

Peine began negotiating with their neighbours in Socaire to gain access to water from a small spring in the hills between the two villages. It was said to have good water and was close enough to Peine to make an infrastructure project feasible, so they registered the spring in Peine's name with the government authorities. Don Humberto remembers that 'the authorities' (the Allende government) promised funding to help build the infrastructure needed to capture the spring and pipe it to Peine. However, the villagers had not yet begun work on the spring when the *Golpe* happened and '*se quedó ahí*' (it all stopped there).[8] What Allende had symbolized, perhaps even promised, to people was not the priority of the new regime.

In material terms, the military regime froze state finance of social projects, including those for basic needs such as Peine's domestic water, and fear of the military paralyzed similar works.[9] The actual violence of the dictatorship also came north. The regime used the abandoned nitrates camps of the early twentieth century in the Atacama as internment camps for political prisoners and the military death squads targeted union leaders and workers in the mines around Calama.[10] Men from Peine working in the copper mines (especially Chuquicamata) and who were union members quit their jobs in fear, some returning to Peine, others to find work in Antofagasta or elsewhere. An early move by Pinochet was also to replace local leaders of the Junta de Vecinos[11] throughout the nation, removing elected councillors and curtailing village (or neighbourhood) meetings, even in remote places such as Peine and the other towns of the Salar. With its reach into the JV and the militarization of territory by Pinochet's regime of the Atacama Desert (see Gundermann 2002), locally appointed leaders refrained from speaking in opposition. By the early 1980s, with some of the immediacy of violence receding, village leaders in Peine decided to return to the spring they had registered a decade before. The leaders found that in the interim their neighbours from Socaire had harnessed the spring waters and begun to cultivate fields around it. According to Chile's Water Code (*Codigo de Aguas* 1981), rights in water must be exercised – the water must be used to retain ownership – and with Peine inactive, members of the Socaire community had reclaimed the use of the spring.

Peine's *asamblea* decided they would find another water source, drawing on their knowledge of the cordillera across which many families had *estancias* (home shelters with animal corral enclosures) in areas of highland pasture. In these they had spent summer months or stopped to rest on their way to other pasture, or during crossings to Argentina sell animals and buy mercantile goods (M. Núñez 2015; Kalazich 2013; Molina 2017).[12] The closest springs, within a radius of approximately 20km from the village, in ravines and canyons which during winter (before the drought) had sustained large flocks, had insufficient flows. Don Humberto, as leader of the JV in the 1980s, oversaw a campaign to send a small group of men journeying into the cordillera. Every time they did so, the men brought jars of spring water back to the *asamblea*, where they were passed around so that each person might consider the water's potential.

In one such meeting, a *señora* from Socaire, who had lived in Peine for many years, suggested they try the spring called Chakizoke near the Lakes Miscanti and the Miñiques peak. Don Lando remembered her saying, 'it is only a small water, but very good', and that she turned to Don Ezequiel (a man from Socaire who had married a Peineña) for confirmation, 'wouldn't you say, *vecino*? Perhaps if it were cleaned out it would produce more water'. Ezequiel and two other men journeyed up to the spring, situated in territory associated with Socaire, and brought down a barrel full of Chakizoke's water. In the different versions of this story I've recorded, the teller often emphasizes how the men brought a whole barrel of water back, enough for everyone to sample some. With the quality, taste and clarity of the water, in Don Lando's words, '*se entusiasmaron la gente*', people became excited. The consensus was to try to capture this water for Peine.

Chakizoke is a considerable distance from Peine and would need a grand effort. Local leaders began '*golpeando puertas*', knocking on doors to solicit support from 'the authorities' – municipal and regional government officers of different government departments in the time of the Pinochet regime – for financial and legal assistance to register this water in the name of the village. Some had their doubts, including 'the authorities', who responded that the *pueblo* was too insignificant to deserve investment of the magnitude proposed. Don Lando recounted, 'When we asked them to help us to bring the water, they said "No, you don't have any rights because you are too few [in number]"'.[13] The village decided to raise the funds needed themselves.

A magazine from San Pedro de Atacama wrote up the oral history of these events under the title, 'From Chaquisoke to Peine: The Feat of Water and the Efforts of a Community'.[14] The by-line ran:

> In the 80s, the pueblo began a series of transformations, many of which Peineños consider significant even today. The beginning of the Peineño summers and the emerging necessity to bring more potable water to the locality are two faces of the same coin. (Yacktamarí c2007: 10)

The transformations that the magazine alludes to are the jolt of modernity that came with the arrival of lithium mining the 1980s. The industry's operations nearby meant local men could get paid work nearby, and men from other parts of Chile came to live in the worker's camp at the bottom of the village. The wage labour and industry saw locals open small businesses to meet worker's demands for snacks and cigarettes. While it is unclear what 'two faces of the same coin' exactly refers to, an equivocal feeling of this period of change also arose from people in Peine who told me their version of the story.

The 'Peineño summers' were vacation activities organized by the United Youth (*Juventud Unida*), initiated by the JV in the summer of 1985. Young men from Peine sometimes played soccer/football against non-local workers from the recently opened lithium mine, though the United Youth also divided themselves (and the village) into two teams who competed against one another in different sports and games, raising money from their supporters in the process. Kalazich's (2013) participatory research on Peineño cultural heritage details how two sporting clubs, Club Deportivo Peine and Club Latorre (the latter founded in the 1930s, the former founded some decades before), divided residents according to the location of their houses, those at the top (Peine) and bottom (Latorre) of the village. Throughout the twentieth century, these clubs organized football games for men and basketball for women, and other recreational events in summer, when children enjoyed summer holidays. The clubs had also arranged *minga*, communal work parties, in which villagers re-roofed neighbours' homes or laboured in agricultural plots for festive and ritual preparations (Kalazich 2013: 187–88, 191, 195). In the 1980s, the Juventud Unida brought the *arribeños* (uptowns) and *bajeños* (downtowns) of the Sporting Clubs back to life in sporting competition for the purposes of modern water infrastructure development.

Mateo and Ivana, siblings who were in their teens in the 1980s, told me about spending long months at a boarding school in Toconao, and coming home during their summer vacations excited at the prospect of raising money for the village and its water infrastructure, to 'work for the water'. As the Yacktamarí journalist wrote:

> From the 7 March 1986, the 'Juventud Unida' made the first payments of money of two summers, $78,700 pesos to the Junta de Vecinos to form a Water Committee and on the 4th June of 1987 they [the JV] elected a board of CAR [Comité de Agua Rural, the Rural Water Committee]. (c2007:10)

Young and old raised money locally and an administrative entity formed to make themselves recognisable to the state (the CAR). Mateo and Ivana's generation talk about this time with pride for what they were able to achieve as the United Youth.

In a reflection of a dominant Christian discourse about the ways modern life and money may be entangled in perceptions of immorality (Harris 1989; Parry and Bloch 1989), Mateo also pinpointed that it was exactly this time which 'was when money began to corrupt people'. The money raised by the youth group is said to have disappeared. Some say it was stolen, others that it was mismanaged or lost on the way to the bank. In my experience, this historical critique is as much turned outwards ('money corrupts') as inwards; reminiscing, some told me, 'people became *egoista*', individualist, as they began to earn money. Raising money for water was thus narrated a moral tale about personal responsibility to labour for oneself (family) and one's neighbours. There was a recurring phrase in the telling of this story; that back then, '*la gente era mas unida*' (the people were more unified). The water story is an important local history that upholds an ideal of solidarity and community entwined in moral concerns about social and economic change.

La pala y la picota: Labour for Freshwater

Chile returned to democratic government in 1989, and the coalition of centre and centre-left parties (*la Concertación*) introduced a raft of liberal reforms in its 'transition to democracy'. The Indigenous Act (*Ley Indígena* No. 19.253 of 1993) set out official recognition of the existence of Indigenous groups in both the south and north of the country and legislated for an Indigenous Development Corporation (CONADI), which became responsible for projects to address Indigenous social inequalities.

A new president of Peine's Junta de Vecinos in 1995 took up the freshwater project with gusto, including registering Peine as an Indigenous Community under the Indigenous Act in the hopes of drawing on funds from CONADI.[15] Before the water could be tapped and piped, a road had to be made to carry vehicles and equipment directly up to the spring's location, as well as clear a path to later lay the pipes. The President of the JJVV asked Don Gerardo to lead the work team, because of his extensive knowledge of the mountains, the broader territory and technical skills. In recounting this, Gerardo explained to me that there is a topographical problem with the spring's location that he was concerned would hamper the downward flow of the piped water, necessitating a pump.

> The lake [Miscanti] is below the spring, but there is a high hill on the other side of the lake, and we had to see whether the [piped] water would pass there or

> not. I used a level, just a carpenter's level . . . to analyse whether it would work. And it looked good! But for CONADI to give us funding we needed a technical plan . . .

While his comments betray frustration at CONADI's distrust of local technical skills, Geraldo gleefully finished the story by saying that SCL's (lithium company) engineer, who agreed to produce the technical plan and lodge it with the government agency, made the same calculations as he had. So, with professional sets of paperwork CONADI made a modest investment (providing five million CLP, roughly 7,500 USD). With this contribution the community purchased lengths of pipe for the work to continue.

An *asamblea* of the JV decided that labour would determine the value of each household's access to freshwater once it arrived. The calculation of thirty days labour to earn a freshwater connection included ongoing access 'free', without any further charges to households. 'One day's labour' was made equivalent to aspects of machinic work. If a man lent a personal vehicle to the work team or fuel for a vehicle, he might earn credit or partial credit.[16] Two notebooks still in the possession of the JV register each labourer's name and the activities of the day's work. The title page of the first notebook declares its intention:

> Notebook of the 'Chakizoke' Spring, Miñiques, Potable Water Project. Deeds and Personnel that worked on the road and others with the date 1st April, 1995. On this date the work began on the Chakizoke spring freshwater road [*el camino del agua dulce vertiente Chakizoke*], guided from Miscanti lake to the spring (signed Secretary of the Junta de Vecinos, Peine).[17]

Adult men and women were organized into work teams (*cuadras*) and they began clearing a road. In the notebooks, the first two days of work included seventeen adult men and one woman with her six-year-old son, plus two men volunteered their vehicles and SCL provided fuel. On the days following, one group worked on making the road and another on excavating the site of the spring and building the reservoir, again, just one woman among them, and 'SCL' cooperated by contributing fuel and a mechanical backhoe. Bit by bit, it took the community just under five months to clear a road for the pipes between Peine and Chakizoke.

After a break to avoid the deepest cold of winter and the busy spring months of planting (*la siembra*) and *limpia de canales* (chapter 1), work began again in November 1995, excavating the ground to lay the pipes. Everyone worked, though some remember working harder than others. Don Benicio, like many other local men, worked shifts in SCL lithium mine,

> When you had your *descanso* [shift days off from the mine] you'd go up there for three days. But . . . for example me, ya? Tomorrow, I start my descanso, *al tiro* [right away] I'd go and work . . . everyone worked for one day, two days or so. I worked one hundred days. Me, in my days off I worked a lot. Others? They hardly worked at all!

Don Ángel, who also worked for SCL said,

> We were the *mano de obra* [labour] for the job. With five million pesos we bought the pipes. That was enough for some kilometres and then we worked *con la pala y picota* [with spade and iron bar] making the trench for the pipes. When we had a day off [from SCL] we would work in the cordillera. Escondida lent a machine to solder the pipes, and the *el red para captar la vertiente* [grid to capture the spring]. We went along like that, making trenches, others soldering pipes . . . for two days or three days [at a time]. . . . we were the ones doing the work, everyone with their spade and iron bar . . . then the other days other groups would work when we were on break. We worked hard. Today the young people don't believe us that we did it . . .

By early 1997, the water had been piped from Chakizoke to the plateau just above the village. They were short about five kilometres of pipe. According to Don Bartolo, 'there the water was running onto the ground. People went up and filled tanks and drums to bring down to their houses'. This was when 'the authorities' (CONADI, the Municipality, the Department of Public Works, DOH) and bigger mining companies 'fell over one another', according to Don Bartolo, to help. Here, the tempo of the story speeds up. Pipe was dug into the ground promptly using the mechanical grader and digger borrowed from one mining company and soldering machine from another helped seal the pipe connections. The municipality and DOH built a holding reservoir (also providing a second head for the water pressure) and laid the village water grid.

A faded billboard declaring water government infrastructure investment could be seen at the entrance to the village until about 2016. But people in Peine emphasize that the freshwater project was a success thanks to *their* hard work. The people of the village swung the '*pala* and *picota*' on the freshwater road, travelled far to knock on the doors of 'the authorities' and transnational companies, and drew on local forms of corporate patronage to get the job done.

The work was driven by desire for good clean water piped into the home – a desire that might be characterized in terms of infrastructural modernity – made possible by communal labour, local technical and territorial knowledge, and cunning to wrangle support from powerful others. The early part of the story is characterized by narrators in terms of state 'abandonment' of the village, and later, storytellers' ambivalent account of CONADI's partial support is a similar kind of critique. In contrast, they received continuous though modest cooperation from SCL, and machinery leant by SQM and Escondida sped up later stages of the job. In Escondida's documentary, cited at the beginning of this chapter, there is a triumphant theme of rural people setting an example in self-help, as prominent in the discourse of neoliberal development.[18] However, Don Bartolo's words reflect a local ethics,

> . . . we are doing this for our family, for our children to come, because we are bringing this vital element at great sacrifice . . . What the people of Peine have is that they don't need people to come and do things for them. We need support! And we will do it.

Bartolo's express ideal of village futures as found in forms of solidarity and autonomy is one that also always relies on some external patronage, provides a sense of the persistent moral vision for local flourishing.

The Logic of Good Water

'*Ahora estamos bien con el agüita* [now we are good with this little water]', Don Lando told me in 2015. 'It is the best water in the area. It is only a little water, but for the *pueblo* it is good'. Peine registered 2.3 litres per second of Chakizoke's water with the DGA. Don Lando's implication was that not only was the water good, but it was *enough* water for their needs. The member of the JV who lent me the Chakizoke notebooks to read and study, said, tapping them with his forefinger before he passed them to me, that he had made *exact accounts* of the effort put in by individuals and households. As above, these enabled the community to assess who provided labour on the project, and thus whose house would receive a connection of freshwater. Others, if they wanted the water connection, had to pay.[19] Despite the agreed value and their accounting methods, contentions have arisen. Don Ángel complained that *jovenes* (young people) and those returning to Peine or rebuilding parts of their lives there after living in the city expected water connected to their house 'for nothing',

> *Jovenes* don't believe us that we did this . . . they want to come here and be given water just like that! The water is here, and we say that you must pay a certain amount, you *must* pay . . . . It's logical that the payment will never be equal to what we did, but rather than working they must pay. What, we are going to allow them to avoid paying and give them water free?

Once freshwater was in Peine's grid, Don Ángel thought that the feat of labour he and his contemporaries achieved had been forgotten. The bookkeeping that describes individual effort became less visible as community leaders changed and notebooks gathered dust on the shelf. Given that those connected to the grid paid no ongoing costs, it seemed to some that Peine indeed enjoyed 'free water'.

Supply to households became controversial as many households themselves changed. Beginning in the early 2000s, and facilitated by good water, a handful of Peine households expanded the buildings on their land to accommodate mine workers or provide meals to the increasing numbers of mine and contract workers associated with lithium extraction. Some constructed boarding houses for tens of men. These locals threw hoses from their home's freshwater connection to the new buildings, thereby enabling the generation of profits. Chakizoke's 2.3 litres per second of flow was increasingly insufficient in relation to daily demand. Conflict among neighbours sometimes flared, with accusations of water overuse and non-payment of access. Whether they ran businesses or not, residents living toward the top of the village and those right at the bottom had the worst time of it: at the top, the pressure was bad, sometimes not enough to get the water flowing through the gas heater, and at times at the bottom, there was simply no water at all. In the summer months, when children and young people were home *en masse* from school and work in the city, the problem, people said, was '*crítico*' (dire).

In addition to internal socioeconomic pressures, freshwater infrastructure needed repair. It was also attracting administrative attention from the state, since MOP had introduced regulations relating to the definition of 'potable' water.[20] Don Alfonso wanted to retire from his position on the Water Committee (CAR), taking the opportunity to do so with the impending regulatory changes and the community needed a new set of members. Moreover, they needed to update their systems; if they could not show that they were ensuring water quality through the new specifics of regulation, then state agencies would take over local water management.

In 2013 I attended an *asamblea* called for the purpose of electing a new water committee. Don Alfonso opened by saying, 'Well, this can be short, just let's have some volunteers and have it done with'. There was an initial silence and then the discussion extended into the early hours of the morning. Amelia, secretary of the Atacameño Community of Peine at the time, opened her statement with a rhetorical question, 'What do we need a Water Committee for? Well, some people', she answered herself, with a tone of accusation, 'might want professionals to take over'. Don Eberardo, a key figure in the story of Chakizoke, also made a long speech. He emphasized the hard work of the previous CAR, but that now they needed a new set of people to take this up, and that the community must support leaders once they had taken on the job. A younger person opined that perhaps, instead of volunteers, those responsible for the work could be paid a salary or even small honorarium for doing the job, but others spoke forcefully against the idea. Adriana said,

> We were never paid in the past and the people of this pueblo did beautiful things, like bringing the water here. As children we got up at five in the morning to *trabajar por el agua* [work for the water] up in the cordillera. There should be no payment of *dirigentes* [office bearers], they should do it for the pueblo.[21]

The discussion continued, branched out and refocused, turning on the same points for some hours as all those in the meeting made comment.

Eventually, the president of the community, apparently exasperated and tired, took up a marker pen, and declared, 'Right! I'm taking names of the young people'. Immediately, people began nominating others by calling out the names of men and women middle-aged and younger, each time met by a clamour of appeals from those nominated or their family members. The list of candidates was reduced in number from the original fourteen to nine (five were found to have legitimate reasons to eschew nomination) and the meeting's organizers took to neatly ripping notepaper into strips. As the leaders watched, *asamblea* members wrote on a strip of paper, signed their names and deposited their paper into a wooden box. Once everyone had thus voted, Don Gerardo rose to pull out pieces of paper one at a time, calling out the name on the slip. The person who received the most votes became president of the water committee, second, treasurer, next secretary and then two general directors. Each elected person rose and spoke briefly to accept the nomination, expressing honour in being named to serve the *pueblo* and acknowledging the responsibility of their new position. Each was applauded, and each new committee member signed the minutes.

The apparent reticence to take on a committee position contrasted with the sentiments expressed by each new committee member in their short acceptance speeches. The community leaders, who that year were 'young' people (in their 30s and 40s), had overseen the election of a water committee of the same generation. They were led by Dolores.

The *asamblea* had considered the possibility that all members of *directiva* might be paid, but such a move conflicted with the longstanding ideal of *'voluntad'* or willingness: everyone with the capacity should at some time undertake service to the pueblo through such positions. Nonetheless, holding *'cargo'* (office, committee position, especially as president or treasurer) had become complicated, expensive and technical. Leaders and committee members must spend hours a month travelling to the city (Calama) for *trámites* (bureaucratic procedures), attending regional meetings with the authorities, or take responsibility for local meetings, which, like the one described here, might stretch into the early hours of the morning while inching toward consensus. Those with a *cargo* have cell phones that ring constantly, '*cuesta mucho*', people explained to me; it is difficult, but it also costs money, personal and social resources to take on elected responsibility.[22] In the case of the CAR, the responsibility of making sure their small supply of water consistently stays connected, clean and arrives in everyone's faucets is a huge task. It involves managing water as a scarce social and hydrological resource that makes possible the shifting definition of a good life in Peine.

Two years later, when I spoke to Dolores about what she did first as president of the water committee, she explained that she had pledged to solve the consistent freshwater problem. Like Don Alfonso's works described in the opening of this chapter, Dolores organized communal and ceremonial labour: work to install a new concrete lid for the reservoir at Chakizoke spring to replace the old, corrugated iron one, and a ritual *convido* in the location. I accompanied the group that undertook the day long work party in which the reservoir was physically cleaned of debris, a formal *convido* to the water and mountains conducted, and the tank sealed with concrete. Afterward, with the water source secured, Dolores announced that Peine's CAR would begin to charge each household a metered usage fee on their water.

Charging a fee for Chakizoke's water altered the value of water labour as configured by the previous generation of leaders. Here I quote at length from a recorded interview with Dolores (in 2014) in which she explained her rationale;

> We had, like a water crisis here. Everyone was complaining that there was no water in their houses . . . And people didn't know what to do . . . And there's always been this question of how to control each *el gasto* [household use], but no-one wanted to *andar en esta marcha blanca* [get this new process going] to charge for water. Before, everyone used the water and no one paid a peso. So, when we got in [as a new committee] we decided to say 'everyone talks about getting going with charging for water, but no-one's done it', *nadie se atrevía* [no one has dared to do it]. So it seemed like the best way to control it and for no-one to complain. And we did it . . . and the problem got better . . . So I think it was a good decision . . . You see, when they connected the potable water, the project came with water gauges for each house. So they were there, every house had its own gauge. But no-one wanted to get the process going . . . So it was just about getting it going . . . There have been no complaints and today there is a lot of water, and a decent account of it. . . . Since we started charging we have not had problems of people telling us they have no water.
>
> SB: How does the Comité use the money collected?

> *Dolores:* For example, the tank cleaning. We have the cleaning of the [grey water] plant that is not working, but anyway we have to treat the used water. We have to pay the person who does the account statements. Eh, we pay, um, yeh, all of the gauges, every month, at the end of the month, they take readings so that afterward a bill can be given, and people pay the bill. We sought people from the village who didn't have any work and the person who applied didn't have any, many more applied, but they already had work. So we gave it to her. So the money stays here. We pay a person from here and so it, like, comes back here, there's a circle. . . . There are also emergencies, such as gauges that are deteriorating, pipes falling apart. . . so all of this needs to be paid from what one pays and is saved [by the committee]. Now we want to [use the money to] make a storage shed and buy materials to have in there. Things that when we have some connection falling apart, we have the stock and we don't have to go to Calama, which is so far away. Now we have an office that is a [shipping] container. We bought computers . . . we bought paper, we always need that . . . We have a . . . photocopier. And we've been doing that, *po'*. We've had a program installed in the computers to create the list of *socios*. So now you don't have to do this in a notebook. Now we revise the page and all the consumers appear, the number of their gauge, whether it's up to date, all that. So that's how we've used the money.

Dolores thus emphasized the way the money paid by each water user stays in the community; there is a 'circle' around which money earned by the CAR enabled them to maintain the water system through cleaning reservoirs, fairly employing local people who had previously not had work, and purchasing modern equipment for keeping good records of household payments. While the process had been possible 'from the beginning', because water gauges had been installed for some time, the community had for many years adhered to the ethic of an unmetered water connection; the value of the original communal work assured the water's flow into each household tap. But change to the quantity used and number of users put this value under pressure. Dolores' rationale for getting people to pay for water sought to keep a sense of local value in the logic of her actions, while adjusting the commons to a managed resource as a way of meeting heightened demand.

While her accounting circle may have small leakages, Dolores underlines how they took care to maintain ownership and management of Chakizoke's water. If the authorities considered the Peine community not to be managing their own water supply effectively, they might install an external contractor as manager. In the words of Amelia, in the meeting described above, this would take away local control of their water. Even with all the reluctance people demonstrated in assuming a position on the CAR a few years before, and even in the confronting process of paying for water that they had previously laboured to have 'free', the idea of outsiders managing their freshwater was something the community refused to countenance.

Dolores and the committee asserted Peine's control over decisions regarding the future and history of freshwater. Their plan to respond to contemporary administrative and economic demands with the circularity of water payments, continued the ethic of maintaining relations with their water, outside state interference. It reaches their homes and bodies in good condition because of the community's communal effort, which as I have narrated here includes their politicking with authorities, corporate actors, and neighbouring communities. Good management of the water is about all the many practices that maintain the relation between Chakizoke and the community, and as such the local knowledge of water quality is as much about lived experience of what is good as its chemical composition.

Good but Not Potable

In 2015, a Regional Government representative responsible for domestic water infrastructure, the SEREMI de Salud, addressed an *asamblea* in Peine.[23] She outlined the terms of the '*Vigilancia Sanitaria*' (Health Monitoring, DS735) law, which from 2010, dictated that to be designated 'potable', any water provided for human consumption must be chlorinated at the minimum percentage of 0.2% to the maximum of 2.0% (the slogan of the department was 'Water for Human Beings').[24] The SEREMI announced to the meeting that Peine must adhere to the law.

Sitting next to me, Doña Inés put her hand on my shoulder and pulled on it to whisper in my ear, '. . . you should tell them that you saw the work we did at Chakizoke, tell them that our water is good, it doesn't need *cloro*'. I whispered back, 'maybe it's better, coming from you'. The SEREMI finished speaking and asked if they had any questions. In a voice conveying that she was about to tell a story of some consequence, Inés turned sideways in her seat to address the *asamblea*. She described how the pueblo's labour meant that they knew they could count on Chakizoke's water. Moreover, two years before, the APR organized the work at Chakizoke to ensure that the contents of the reservoir in the cordillera would be safe from contamination by *animalitos* (little animals), or debris from surface flows after rains or snowmelt. The water, Inés went on to describe, emerges from the ground high in the mountains and then, thanks to their work, is captured by a nice clean concrete holding tank. She said, 'we know our water is pure, clear, and beautiful'. Inés then swivelled to address the SEREMI, saying: '*Señorita, la tengo una pregunta. ¿Qué pasaría si claramente no lo hacemos?'* ('Miss, I have a question. What would happen if we simply did not do it [add chlorine]?'). There was uncomfortable laughter among the assembled. The SEREMI paused in the face of the older woman's gentle resistance, and stated that, as an Indigenous community they have a 'right to their culture', but that in this case, they might 'put aside their culture for the good of their health'.

Dolores, as President of the newly renamed APR, stood to read from the minutes of a meeting held some months before. She stated that there was community consensus against chlorination of their water. Moreover, as '*tia*' [Aunt] had just said, over the past two years Peine's APR had overseen the cleaning of the reservoir at the top of the village, as well as repairs to the reservoir at the spring. The SEREMI, nonplussed, replied to Dolores that, 'what you will have to do is accept and adapt to chlorine'. She repeated the legal obligation of the APR to add chlorine to their drinking water.

Three members of the *asamblea* spoke in a clamour; one complaining that the SEREMI's attitude showed a lack of respect for the principles of Convention 169 of the International Labor Organization, and Dolores reiterated that the community had reached consensus against chlorination. Don Gerardo stood and took an oratorical tone. He said the *pueblo* has autonomy in this matter given his and others' knowledge about the water itself, its origins, and association with the Miñiques peak, and the community's labour associated with the pipes, tanks and other infrastructure. He emphasized their practical and technical knowledge, and he pointed to the SEREMI and said, 'not just culture', and proceeded to offer a short account of government abandonment. In the past, he noted, the authorities had refused to assist Peine when they sought support for bringing this source of water; instead, they had received some help from mining companies. Then, when the authorities finally agreed to help by laying pipes in the town, it failed to make sure contractors installed them effectively. 'There are still problems!' he exclaimed, and members of the assembly expressed noisy agreement. Faced with their claims to history, the SEREMI valiantly replied that in her opinion, 'there are more positives than negatives' to water chlorination.

Don Gerardo turned to point to a government employee who in 2011, had worked for DOH, overseeing the contractors working on the village water grid. Using the respectful honorific to show that he had been the boss of work, and his name ('Don Alvaro'), Gerardo asked, 'can you help us with this?' Looking uncomfortable, Alvaro replied that this was the first case of chlorine refusal in the region, but he agreed to find out if it were possible for the village to decline and still adhere to the law. Gerardo retorted that what was missing from the DOH's work on the water system thus far were many technical details. Local infrastructure and projects financed by the government agency and begun by contractors had never been completed and he thought never would be. He said, '*si nosotros tenemos que cumplir, tambien ustedes cumplen*' (if we have to fulfil our obligations, then you must also fulfil yours). Alvaro responded apologetically and recommended that they take their refusal to a higher level of government than his position allowed. The SEREMI left the meeting.

While the *asamblea* refused the addition of chlorine in their freshwater from Chakizoke in 2015, this is not a simple story of resistance to the state. Chlorine's bacteria-eradicating effects are part of everyday knowledge of sanitation processes. It has been common in Peine, and throughout Chile (including metropolitan areas) to have a bottle of 'Cloro' (a commercial brand) under the kitchen sink. One puts a little into the water with detergent to wash dishes, may use a few drops in the basin of rinse water, and I was instructed to squirt it straight from the bottle all over the floor before mopping. I was admonished for not wearing rubber gloves when washing dishes in water that contained *cloro*.[25] After this meeting about *cloro*, Inés, her adult son and I were sitting in her kitchen with a cup of tea, and she talked further about her distaste for the idea of chlorinating the water and giggled at her courage to speak against the SEREMI. 'If they put *cloro* in it, the water would have "*otro sabor*" (another flavour)', Inés lamented. Peine's consensus against adding chlorine was not a fear of a new or an unknown danger to which a remote Indigenous community must adapt, as the SEREMI said, but their intimate knowledge of what is good and bad in practical, qualitative, and sensorial terms. That other flavour Inés evoked was not only the reliable poison of a familiar disinfectant, but intervention in their good water from Chakizoke, water that need not be improved since it had already been secured with physical, social, economic and ritual labour.

The improvement of sanitary regulations in relation to rural water supplies defined potable water as chlorinated and while this law undeniably sought to address a national problem of access to good drinking water in rural areas of Chile, in this case its application ignored the way in which Chakizoke, as the source of Peine's freshwater, was locally ensured as good for local life. State approaches to water provision, and actions to create relations with Indigenous communities about their water through consultation (such as the SEREMI's visit) only partially correspond with community understandings regarding the ways that water may be governed, what constitutes good water, and how such judgements can be made.

Conclusion

From where it bubbles up out of the ground, Chakizoke's water is directed to a tiny reservoir, then enters pipes and flows mostly downhill for about 45 kilometres without the aid of mechanical pumps. When it reaches the uppermost limits of Peine, the pipes enter a stone and concrete reservoir with a sealed top, the tank protected from human and animal ingress by a high fence. Water exits the tank directly into the pipes which form the beginning of the water grid. According to Dolores and others, tests of their freshwater have shown that it is the purest water in the region. There exist many other such sources of high altitude, very clean, spring water, but Peine community members and their guests enjoyed water of this quality in their home taps which was free of both naturally occurring as well as mining pollutant-produced arsenic, and for many years, also free of chlorine.[26]

In the late 1990s, with Chakizoke secured, freshwater modernity, and its excesses, flowed from Peine's household taps. However, keeping the freshwater exiting Peine's taps equally and freely became difficult to sustain at the original labour value of common water provision to all households. The water attracted young people back from the city during summer vacations and nourished people's development desires, including those of local business owners who were able to offer good water to labourers and mining contractors, and in turn make a profit from accommodation and catering. The younger generation of village leaders found ways to keep the substance of the freshwater circulating within the community, by charging residents by volume of usage, but retained local control of both water management and any proceeds from water fees, and thus the relation between Chakizoke and the residents of Peine.

The earlier events narrated in this chapter, that some in Peine refer to as the 'freshwater road', intersect with the years in which Minera Escondida ramped up groundwater extraction in the southern Salar, as well as those in which the Chilean state officially recognized Indigenous peoples. It was also a time of shift in global extractive industry relations with Indigenous Peoples, when some of the biggest players in mining (among them BHP, Escondida's operating company) began to introduce social and environmental policy and seek 'socially responsible' relations with so-called 'host' communities. The actors that Peine community members mention as key to the success of the freshwater road are local people, especially the strong hand of community leaders who kept workers in line and happy during the two years of hard work to finish the project. The collaboration of lithium mine (SCL) bosses always receives a mention when locals told me this story, since they too were neighbours and they also depended on

Figure 3.3. Chakizoke spring, Miscanti Lake in the background. © Sally Babidge.

local freshwater. While Escondida's lending of a soldering machine for the pipes was often mentioned in the telling, most left implicit an important historical conjuncture: the connection of the Chakizoke's water to Peine in 1997, was the same year that the Peine community signed the first contractual agreement in Chile between an Indigenous community and a mining company for company-provided 'development benefits'. Had good water coincided with good company?

In the late 1990s, Peine had a source of freshwater on the near horizon and was recognized by the state as Indigenous people with rights to claim territory and waters. They had formulated a local development project and relied on their own expertise and calculations, undertaking technical studies of topography and flow, apportioning communal labour, and relied on a willingness to undertake communal labour to make their own 'development project' a reality. To do so, the community also sought funding from the state and patronage from powerful mining companies, which they characterized as 'collaboration'. The Chakisoke project marked the beginning of their first formal relationships with transnational capital. The following chapters examine the dynamic of these relations.

Notes

1. Lit. 'When they put in the batteries . . .' '*Ponerse la pila*'. Chilean colloquial phrase meaning to get fired up, get oneself together.
2. An IBC is 'Intermediate Bulk Container', a 1000 litre plastic tank stabilized by a steel frame.
3. In 2013, SCL was still purchasing domestic (non-industrial use) water by the truckload from various other sources.
4. For example, in San Pedro de Atacama they also used irrigation water for drinking and were seeking better water, as explored in a documentary piece by Chilean filmmaker Omar Villegas (1987) in '*Atacameños: Una Nacion Que Resiste*'.
5. In 2017, during a participatory mapping research workshop in Peine to consider local understanding of environmental change related to water extraction on the Salar (Babidge et al. 2019) the arrival of freshwater and its infrastructure in the late 1990s in Peine was nominated by one person as one of the most momentous 'environmental' changes in Peine's history.
6. The CAR was a locally elected committee whose members took responsibility for ensuring good water to the community and administratively dealt with the state agencies of DOH, DGA and MOP. It later became the Agua Potable Rural (APR), see below.
7. Chakizoke is in the Community of Socaire's territory. Socaire agreed to Peine's capture of a specific volume of its waters, which the Community of Peine registered with the DGA as a concession, not as traditional waters.
8. *Golpe del Estado* – the Military Coup in 1973 led by General Augusto Pinochet, then head of the Chilean armed forces. Former President Salvador Allende died from gunshot wounds during the September 11 coup d'état. Chapter 5 picks up different parts of this history.
9. There is a large volume of literature that focusses on this time by Chilean social historians and others.
10. Guzmán explores the political, physical and memorial landscapes of this legacy in his film *Nostalgia de la Luz* (Nostalgia for the light) (2010). Vilches (2011) takes a memory and material culture approach to the history and architecture of these encampments.

11. The law of the JV was enacted in 1927 to bring together elected leadership in neighborhoods and small towns and make them accountable to the Chilean state. It is thus a well-established institution of the lowest level of the Chilean state, with a national history of political activism.
12. While demanding the crossing of the Andean cordillera, Salta (Argentina) was historically preferred to Calama (Chile), since the road from San Pedro to Calama is '*pleno desierto*' (over 100 kilometres of mostly rock and sand, desert plain) with few watered stops for the herd. A history of extensive territorial use through herding and has been explored in detailed by those named authors.
13. Direct quote, '*Cuando pedimos que nos ayudara pa' traer el agüita, nos dijeron que "no, ustedes no tienen derechos porque son muy pocos"*'.
14. Aside from citations of other authors' work, I use the spelling '*Chakizoke*', following the Junta de Vecinos records.
15. As it was explained to me, the President of the JV at the time formed the Indigenous Community exactly in order to access funds newly available from the state body CONADI. It is likely that there were various impulses for forming a registered community, especially lodging territorial claims for Indigenous title to lands and waters.
16. At this time, as far as I know, no women were owners of vehicles.
17. While the Peine Indigenous Community had been registered with CONADI by this time, a corporate body separate from the Junta de Vecinos was not yet operating locally.
18. The quote from the film's narrator may also be read for its pejorative connotations, as Indigenous peoples the stereotype was as inconsistent workers.
19. If a person had a shortfall of days, they were asked to make up the difference in cash payment, and those who did not labor at all during the freshwater road had to pay a one-off amount of roughly 900,000 Chilean pesos.
20. The committee gained a new title 'Agua Potable Rural'(APR). The changes had particular implications for SCL (by 2013, operating as Rockwood), who must provide their workers with (legally defined) potable water.
21. As elsewhere, I have left pueblo untranslated to indicate the apparent meaning; that 'pueblo' can be the place, but also 'el pueblo' is 'the people' (with the socio-political intent this implies).
22. Chapter 7 takes up leaders' complaint about the '*sobrecarga*', the overburden, of mining projects to which they were responding. I have translated 'cargo' here as 'committee position', though the parallels of taking office and carrying a burden are relevant to the increasing weight of these roles.
23. *Secretaria Regional Ministerial* or SEREMI is a regional representative of a Chilean central government ministry, in this case of the Health Department. Each of these regional secretaries of different ministries is simply referred to in local parlance as "the SEREMI" (if more than one present, then their ministry is added) and I have adopted this way of referring here.
24. The law is available at: https://www.leychile.cl/Navegar?idNorma=197226, last accessed 13 April 2017).
25. Chlorination of kitchen sink water is sometimes explained by the fact that hot water is not commonly used in to wash dishes (lighting the gas water heater is saved for bathing).
26. In December 2019, Dolores told me that the CAR had finally agreed to chlorinate the water, but that it was only the absolute minimum. She asked me, 'can you sense it?' There was just the faintest hint of *cloro*.

Chapter 4

AGREEMENTS, 'DEVELOPMENT BENEFITS' AND THEIR MORAL ECONOMIES

The first time we spoke, Miguel said to me that 'when Convention 169 is ratified this year . . . we will be able to really sit down and discuss with the companies, that is, as befitting of our status, at one table'. For Miguel and other Atacameño leaders I spoke to in the first two or three years of field research, '169' was on everyone's lips. Ratification of Convention 169 of the International Labor Organization (C169, ILO) was promised by the first *Constitución* government at the time of return to democracy but had taken some fourteen years to fulfil. I asked Miguel what these changes might entail, and he explained his hopes as I partly quote above; that when C169 was ratified, they would gain ground with mining companies, because this global instrument of recognition of their status as Indigenous peoples would influence national laws, giving them a 'seat at the table'.

C169 of the ILO sets out Indigenous Peoples' rights to self-determination, as well as state signatories' legal responsibility to promote those rights. An Atacameño councillor for the state agency body CONADI explained to me that for him, it represented 'a form of respect. . . . that we are persons here. We belong to this land (*tierra*), and we need that to be respected'. Earlier recognition of Atacameño peoples as an *'etnia'* (ethnic group) under Chile's Indigenous Act (1993), led to community members such as those in Peine registering an Atacameño Community which enabled engagement of the state and mining companies such as Escondida and associated leveraging of funds for basic infrastructure or other activities. When Chile became a signatory to C169, it promised heightened forms of recognition of Atacameño peoples, with the potential to deliver what the earlier Indigenous Act had not. In particular, greater authority over territory.

Notes for this chapter begin on page 131.

That first time we spoke, Miguel recounted,

> We have been dealing with Minera Escondida, who produce, who are going to produce, extensive environmental damage in the Tilopozo sector. They take water some kilometres above there, *sacan, chupan* [they suck it out of there],[1] for the company. And we have with them a development fund to undertake various social activities. But it hasn't been good from the beginning. We had to sit down with them, from time to time it was really tense. It was not easy. That is, it is difficult to make the company understand much, even now. But we are on that road, because we have the recognition of territory, and they are inserted inside that territory.

In 1997, the president of a newly incorporated Atacameño Community of Peine signed an agreement with BHP Billiton, the multinational company (now BHP) that operates the Minera Escondida copper mine. Doña Valeria remembered '*un rubio*', a blonde guy, then Australian CEO, came to Peine to sign the contract. Ten years later, the community and company signed a new agreement after drawn out negotiations. The development fund Miguel refers to is the focus of both agreements, which delimit the amount that Escondida must pay annually into a trust account and the terms by which Atacameño Community of Peine can use these funds. The community ostensibly gains a share of the 'development benefits' from the copper mine.

Neither the first nor the second agreement makes explicit mention of Indigenous territorial rights or the environmental or social impacts of company groundwater extraction or effects of that extraction for territorial rights. The territorial effects of the mine and associated groundwater extraction is at the heart of the relationship between Escondida and the Indigenous community, as a manager of Escondida community relations also recognized (in an interview in 2011), though in less explicit terms than Manuel:

> In our [Escondida's] neighbourhood is Peine. They are not a direct proximity neighbour, but [a neighbour] in principle and according to the effect produced by our extraction of water, in the Salar and to Tilopozo, which is a pasture area very close to Peine. This is *lo que fija* [what determines] our relationship.

For the manager, company water extraction associated with mineral processing occurs 'very close to Peine', while the Escondida mine is over 100 kilometres southwest from the village, high in the dusty coastal plateau hidden from most people's everyday life. The manager's explanation of the basis of their relationship, as we sat in his office in company headquarters in Antofagasta, was as 'neighbours'. Both the closeness of Tilopozo and the distance of Peine from the mine ('not a direct proximity neighbour') aligns carefully with the way development 'impact' and human residence is calculated in extractive logics. It emphasizes only one environment (Tilopozo, a 'pasture area') in the Salar, as though the place is alone in the landscape. An important contrast, in semantics and value, is evident between the manager's statement and Miguel's, in which the latter emphasized that the company is 'inside [Peine's] territory', whereas the manager notes the community is 'in [the company's] neighbourhood'. The contrast marks a profound disagreement between the community and company which is also the basis of their agreement-making.

This chapter examines the social history of contractual agreement-making between Peine and Escondida. It considers agreements as legal devices that have arisen from a global liberal imaginary, and given the limitations of state regulatory function, circulate and are transformed as they interact with particular sets of industrial and legal relationships and the ecological and economic terms of locally conceived development. As instruments, agreements, like C169, belong neither wholly to universalizing liberal and capitalist imaginaries nor can they be imagined arising from emplaced contexts (Tsing 2005). Instead in quite different ways, each agreement as a politico-legal and technical instrument is at once both a legal abstract and part of the world in which they operate. As documents they represent a formalised relationship between disparate groups (miners, Indigenous Peoples), and make licit certain territorializations. In real terms, the Peine-Escondida (community-company) relationship is only partially described and delimited in the agreement documents, but they are a material focus for both parties' development aspirations for the community. Industry and critical literature has examined the expanding phenomena of agreements in Latin America and on the reasons why companies enter into agreements with Indigenous communities; as an instrument of extractive governance, a company may seek an agreement in order to set the terms of relations, to minimize conflict that may arise due to industrial production of environmental damage or social tensions (e.g. Bustamante and Martin 2018; Odell 2021). But why might an Indigenous community seek this measure?

An agreement between a company and an Indigenous community is late capitalist extractive governance, in which transnational capital devises ever-new instruments and discourse to respond to their critics (Kirsch 2014; Goodale 2016) in line with norms imagined by global civil society (Garsten and Jacobsson 2011; Falleti and Riofrancos 2018). Order is imagined to be maintained. But agreements are not only this. The ethnographic material I narrate here relies on explanations and political analyses put forward by leaders and ex-presidents of the Atacameño Community of Peine, who represented agreement-making with mining giants as a way to expand their claims to territory, and thus to state and company recognition of their authority in matters territorial. Agreements, as documents, are central to the material politics of advanced extractivism. They are documentary mechanisms that the community imagines will help secure some terms of an unequal and insecure relation.

Agreements may also be read in terms of an older materialist analysis, as articulating the differential terms of the moral economy of development. Given these matters, this chapter first considers what agreements are and the moral economy by which the effects and processes of the Peine-Escondida agreements may be analysed. In the conditions of Chile's advanced extractivism, Indigenous claims to territory and waters have been more expansively exercised, and in parallel (rather than at cross-purposes) transnational capital has undertaken new and expanded mineral and water extraction.[2] I begin by tracing how community-company agreements are part of these dynamics of recognition of Indigenous rights and apparent contradiction of expanding extraction from Indigenous territories. In the second half of the chapter, a focus on community efforts to determine the conditions of their relations with Escondida reveals how their relations exceed the terms of contractual agreements as signed. Peine's 'Development Plan' is a record of the structural realities and cultural and economic aspirations, which were rendered in the form of discrete projects that could receive the 'benefits' from the Peine-Escondida agreement. The Plan reveals how the agreement process had a hand in shaping local futures, as well as the ecological futures imagined by the community that go beyond the 'development' that the company agreed to fund.

What Is an Agreement?

Indigenous community-mining company agreements like those signed between Peine and Escondida have been a global phenomenon since the 1970s, commonplace in Australia and Canada, but until the 2010s were rare in Chile (O'Fairchealleagh 2016). A range of critical literature situates these merely as instruments of neoliberalism, a divestment of state responsibility, and a technology by which companies secure community acquiescence to extractive activity (O'Fairchealleagh and Babidge 2023). According to this logic, agreement making in which mining company provision of monies and infrastructure, such as football stadiums and water infrastructure and supply to Indigenous communities, has the effect of conveying achievement of company 'social license' for company extractive activities (Owen and Kemp 2012; Prno and Slocombe 2012). Communities, while gaining some economic benefits, are imagined to trade-off their territorial rights in agreements. Agreements thus attract negative political and critical attention, indeed; communities who engage in agreement making with extractive interests may be criticized by other Indigenous peoples or environmental activists as having 'sold out'.[3] In the logic of this critique, signing an agreement for development benefits situates communities as (relatively powerless) agents of neoliberal rationality in which the entrepreneurial 'self' enacts economic ends (Gershon 2011). Instead, I argue that agreements may be diverse social and material phenomena in origin and ethical intention, and articulate some dynamics of Atacameñoagency.

This chapter is not a technical evaluation of the Peine Community-Escondida company agreements, in which sections of the contract agreements are cited and analysed for their fairness, for example, or potential effects. During different periods of fieldwork, interviews and conversations, I have read the Peine-Escondida agreement documents, witnessed their terms play out in community meetings in Peine, with and without company employees present, and, during one conversation, had a copy of one agreement document waved at me theatrically. Two questions I continued to ask over the decade or so of my research in Peine and the broader region was, 'how did members of the Peine community experience a relationship with Escondida (as a result of their formal agreement)?' and 'what were the development benefits actually distributed by extractive industry?' Judgement about the value, or distributive successes and failures of 'development benefits' were central to people's discussion with me about their experience of the relation generated by the community agreement with Escondida. Thus, the chapter is concerned as much with the events and imaginaries of 'development' leading up to and during the signing of agreement, and the broader relationships of extractive development, as it is an exploration of these particular agreement operations.

Industry cheerleaders for community-company agreements see them as instruments for a conflict free industrial landscape and a means by which industry can structure their claims to improve life in impoverished communities. Critics of the instrument of agreements sometimes assume that the economic transactions outlined in agreements and that result in benefits to communities, generate a community obligation to accede to company requests or proposals, especially to accept the equivalence of a trade-off of such benefits for environmental damage. In this chapter, I show how the assumption that agreement relations are transactional (financial benefits for acquiescence to industrial use of territory) elides the fact that parties to an agreement may well understand them on their own terms, for example, they may not be envisaged as a transaction at all.

Rather than examining agreements as defined by economic transaction, I consider them as 'moral economy', a form of *social* exchange.

In philosopher of science Lorraine Dalston's theorization of the moral economy, it is a 'network of values charged with affects' that are 'historically created, modified and destroyed' (Daston 1995: 4, 7). In her usage, the moral economy refers 'not to money, markets, labour, production, and distribution of material resources, but rather to an organized system that displays certain regularities, regularities that are explicable but not always predictable in their detail' (ibid.: 4). This broad focus on value that arises from social practice, or the cultural evaluations of just practice, parallels anthropologists' use of the term (see Fassin 2011; Beresford et al. 2023). The evaluation of commitments made legitimate through practice are a substantivist argument for the mutual obligations among persons (and groups) that arise from social and material transactions between them over time (Carrier 2018, see also Trawick 2001; Rajak 2011a; Gregory and Altman 2018). The moral economy is an analytic for considering the terms and performance of obligation between companies and communities as parties to an agreement, and for examining the dynamics of people's evaluation of associated 'relationships and their histories' (Carrier 2018, p.32). If not acquiescence to one another's expectations, what do agreements, as formal contracts sought out by both Atacameño community leaders and mining companies, generate? Diverging values and relational futures may be traced to the event of the Peine community and Escondida making agreements. I turn now to consider accounts of agreement negotiations and resulting social relationships that were implicated in, contingent on or in explicit rejection of contractual terms.

The State, Territoriality and Development

Before the Indigenous Act, Atacameño-Likanantay people had seen their territories and waters granted by the National lands department (*Bienes Nacionales*) to corporations for resource extraction. As Doña Valeria said '*eramos como cualquier otro Chileno*' (we were like any other Chilean), with no specific rights to the broader territories that they and their ancestors had lived in, moved across and cared for. The lithium mines and copper mines each appeared, installed, and extracted.

The history of the Escondida agreement is temporally bound with Doña Valeria's time as a community leader, she served as Secretary to the first Community President in the late 1990s and was elected President of the Indigenous Community of Peine in the early 2000s. These were during the first years of Chilean state legislation which nominally recognized

Atacameño peoples as holding some territorial rights, and in hindsight, she reflected that they were not wholly aware of the implications of the changes.

> Well, it was like we were so . . . well, ignorant, innocent of all these things that the mining companies [did] . . . of agreements, and well, we used the lands for a long time, we knew that we were using these waters and lands and everything but . . . They were never considered ours until the law came out. Then we became knowledgeable about the law. Only then did we find out that it belonged to us. The lands, the water, belong to us. In this time, Escondida came to talk with us and that was when we made the agreement, more or less, the first agreement that was done.

The Peine-Escondida agreements were the first and early examples of their kind in Chile. The sociolegal phenomenon was new for both parties and the agreements themselves were novel as were the exercise of Indigenous rights. When the Community first signed the agreement with Escondida, as Miguel said, 'the *pueblo* didn't necessarily understand the extent of their rights . . . we know the places we had for pasturing animals, where our places were in the countryside, and that these places belonged to us, but *más juridicamente, mas allá* [the juridical nature of this] was not acted on'.

The Indigenous Act had made provision for superficial waters in northern desert regions to be inalienable if registered within lands owned in the name of a titled Indigenous Community. The water provisions were particularly relevant for Atacameño people who were able to register the water ownership (more correctly, 'regularize' previous use rights) of those waters they used for agricultural, domestic and herding purposes.[4] In the village of Peine territorial and water registration under the Indigenous Act meant communal title for water flows that appear as springs in the village (Vilti and Puri), as well as the spring (Tarajne) that feeds the agricultural oasis of Tilomonte. In 1997, the Chilean Water Authority (DGA) and the National Indigenous Development Corporation (CONADI) signed a memorandum to 'protect, constitute and re-establish' water rights for Indigenous groups in the northern desert regions (Castro-Lucic 2002: 198). Rights to flows of water that feed pasture and thus support herding were registered under the interdepartmental agreement between the Water Authority (DGA) and the Indigenous development corporation (CONADI 1997). However, in the Salar de Atacama and elsewhere, Indigenous people have had less success in advancing their claims (e.g. Lemus 2019). Comparative legal scholar MacPherson (2017) explains how Indigenous water rights in Chilean law rely on demonstrative exercise of rights, for example, infrastructure and improvements (especially irrigation canals and reservoirs), according to 'ancestral use' of resources (Article 64 of Chile's Indigenous Act). In the broader Atacameño territory, herders' pastured animals, also only intermittent toward the end of the twentieth century, leave fewer traces than in the irrigation of crops in villages such as Peine. Moreover, the installation of lithium and copper mining infrastructure delimited people's access and use of lands and waters.

Parts of the southern Salar de Atacama, the canyon and plateau stretching south, the Salar de Punta Negra, lands around Llullaillaco peak and lands between there and Pular peak, including named paths or roads that link these were identified under Peine's communal tenure (M. Núñez 2015). This territorial custodianship is internally differentiated. It includes communal territory and family-specific property in pasture outside the village (which is incomplete), as well as individually owned agricultural plots. Farmed plots of land in the oases of Peine and Tilomonte are the lived sectors of the communal title to Indigenous territory, while also being the inherited property of individual men and women (see Chapter 1). The broader 'Peine Community' territory stretches from the Llullaillaco peak in the south, to the southern Salar de Atacama and village surrounds and up into the cordillera, includes sectors of pasture contain herder shelters and *estancias* (substantial dwellings, often with adjoining corrals) in highland pasture that are identified with individual *family* histories. Peine's territorial claim, like the neighbouring community Socaire's, is a geospatial description of customary herding, pastural and ritual practice and the paths that link place in which such practice occurs. Boundaries may be indicated by *apachetas*, stone markers, in some places, especially in the high mountain pasture.[5] Valeria's expression '*cualquier otro chileno*' (any other Chilean) elaborates on the fact that until the 1990s, Chile had not recognized Indigenous rights to exercise Indigenous land ownership, though her comment proposes a broader critique; while the Indigenous Act stated that territory and waters 'belong to [them]', it did little to practically ensure the enactment and exercise of those rights as against others, especially extractive capital.

Don Bartolo emphasized a cartographic dynamic to extractive effects, unrecognized by any form of state regulation. Escondida's activities south of the Salar de Atacama changed the names of places that Bartolo had always known as Peine's territories, especially in the area the company (and the neighbour company Zaldívar) refer to as Monturaqui and Negrillar. In parallel to what West (2016) calls the 'semantic dispossession' of Indigenous territories, the renaming of these remote, but locally well-known places are apparently small acts that can be radical reterritorializations. Bartolo insists instead that the rocky pass through black hills is 'Mal Paso', and from there you pass through 'Honduras', 'Aparejado', and an area called Pingüeal, where the small robust bush

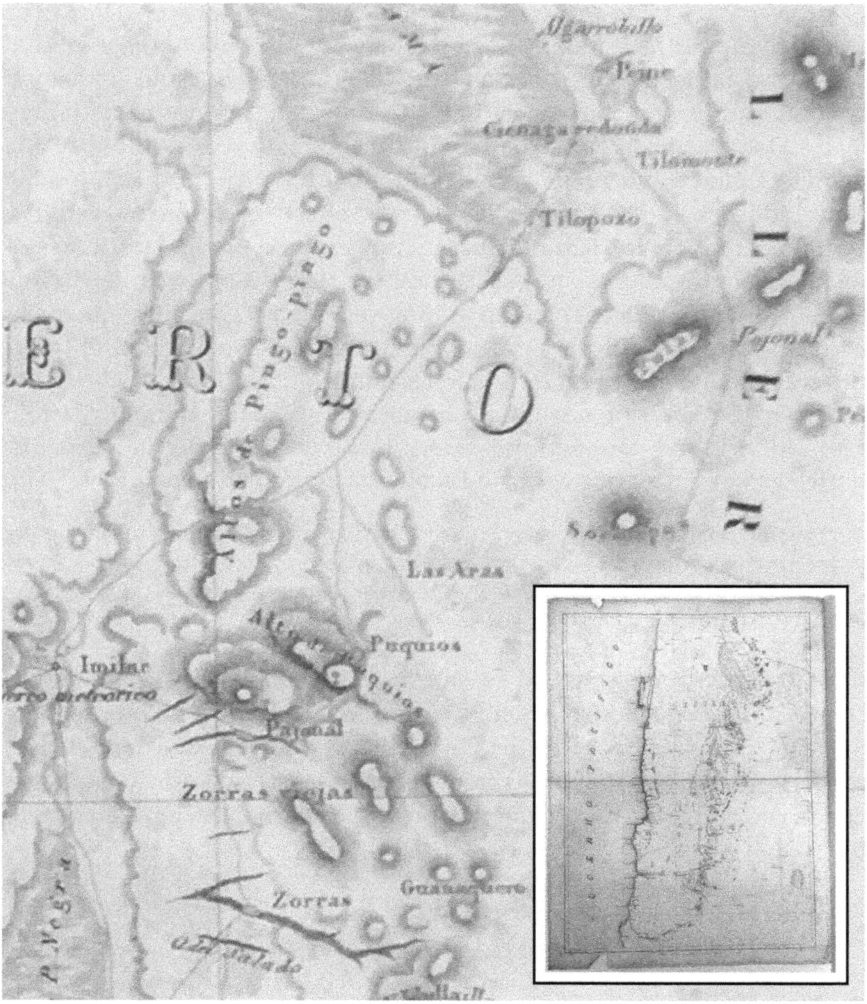

Figure 4.1. Detail of geological survey map (Mahlman 1854) showing places mentioned by Don Bartolo (copy of original held by the Biblioteca Nacional Mariano Moreno de la República Argentina, BNMM).

known as *pingo-pingo* (*ephedra chilensis*) grows. These toponyms describe places known according to a pace of foot, mule or bicycle travel and a dependence on ecology and landscape as pastoralists, artisanal miners (*pirquiñeros*) or workers at the railway in Monturaqui before the vehicular traffic of the industrial mining era (see figure 4.1). In the past there were just two roads (*caminos troperos*, or walking tracks) to Monturaqui, the one called Honduras, and the other Bartolo referred to as Tambillo, the latter passing close to a section of the now World Heritage listed

Qapac Ñan (Inca Road). Since both Escondida and Zaldívar installed their water pumping and monitoring, workers for the mines have graded roads and rallied along countless dirt tracks, damaging old paths and creating confusion about where in the territory one or another road may lead.[6] Don Bartolo's authoritative and performative recounting of the names of places and roads lays claim to that territory, and at the same time indicates one of the less spectacular effects of extractive activity. While renaming and road making are not counted in regulatory assessments of environmental and social 'impact', landscapes with few roads become intensively marked with mining vehicular tracks in the course of extractive projects. Efforts to speak the names of these places, and travel through them knowingly, resist such erasures.

Given the scale of territorial change generated by mining, Bartolo asserted that his community leaders must ensure that the 'benefits' derived from the agreement with Escondida are something tangible:

> It should not just be used to make the pueblo look pretty [*pintoresco*]. The agreement with Escondida . . . this has to have something in exchange. The agreement is a help for us in exchange for the damage they are doing to the territory, the contamination of the mine.

Bartolo and others struggled to ensure they gained something of substance in their relationship with Escondida. This was a relation people in Peine represent as generated by their own interest but nonetheless one plagued by two material absences in the relations: knowledge of the effects of company groundwater extraction and any protective actions of the state.

'We Did It Ourselves'

'*Lo hicimos nosotros*' (we did it ourselves), Doña Rosa corrected me, when I asked her about how the company approached the community to sign a formal agreement contract for development benefits. Gaining Escondida's assistance in the freshwater project (Chakizoke spring) was the community leaders' idea, Rosa explained, and the approach led to the company and community signing a formal agreement. Reflecting its origin in the water infrastructure project, the agreement they signed in 1997 outlined the creation and administration of a 'Development Fund'. The annual amount of 'development benefits' to be paid by the company (for a period of fifteen years) was a little less than 50,000 USD. The Fund was to be administered by a *Comité de Adelantado* (Forward Planning Committee) comprised of representatives of Escondida's social relations employees, the Director of Escondida's philanthropic foundation, the Minera Escondida Foundation (FME), an older man from the Peine community, nominated for his skills and interests in business, and a representative of the state agency CONADI. FME, the Director informed me in 2009 (in an interview held in the FME San Pedro offices), operated under a principle of achieving 'development of capacity in people and institutions'.

With funding from the first agreement, as well as from the CONADI government programs at the time, some small projects were initiated in Peine in the early 2000s. For example, local men were employed to improve the *Ckunsa Turi* Guest House, which gained additional rooms, had its bathroom facilities renovated, and a façade of quarried stone on the masonry bricks, adding quality and 'character' to the buildings designed to please a foreign tourist market. Funds were spent on improving stretches of irrigation ditches with some canalized with concrete. The village primary school received new teachers' quarters with ensuites. Construction began on the Llullaillaco Stadium. Peine's swimming pool, which increasingly attracted visitors from other villages on the weekends, and some of the more adventurous tourists from San Pedro de Atacama, gained designs for the part construction of a new stone and wooden boundary fence and changerooms. These incremental improvements in the village infrastructure were projects and ideas that community members spoke to me about as having initiated *themselves*, using the funding they captured from the company through their own hard work. Some may reflect Don Bartolo's criticism about the money only being enough to 'pretty-up' the village, but the works also articulated, in material terms, a future to which some people aspired; tourism as an alternative to mining.

When she was president of the community, Doña Valeria led some of these improvements. Under her leadership, the *asamblea* decided that the first agreement made with Escondida was an insufficient measure of the value of Escondida and Zaldívar's water extraction from their territories. She told me, 'the money was a very small amount and the water they were using was a lot'. The community leaders embarked on a process of renegotiation. At around this time, the company invited all community members to take a tour of the mine and the areas where Escondida and neighbouring copper mine Zaldívar extract water. Valeria told me,

> The company invited us and all the people went to Escondida; we looked at the wells, we saw the mine, and we went to the Zaldívar mine as well. They paid for the buses and all that. We also went to Llullaillaco and then to Monturaqui, where they take water, and we went with [Don Geraldo][7] there because he knows that area.

Unlike the lithium mines in the southern Salar, Escondida's mode of operation is largely available only in company reporting, and tours like the one the company offered were organized so the community might 'get to know' the mine and its operations. It was a chance for the company to influence the community with the importance of their enterprise as a technical and professional achievement. It was indeed impressive, and Peine community members encouraged me to go and see the extent of Escondida's infrastructure.

Activities like the tour, Valeria explained, were organized in the time of a social relations boss who was friendly and good to deal with. She said,

> A *caballero* called Ojeda was the contact from Escondida,[8] he was *amable, muy, muy, muy* [very friendly] and said 'let's sit down and talk about what is possible'. Then he left the company and a new man entered . . . That, then, was the beginning of a big fight. We went looking for a lawyer and with the lawyer fighting against them and them against us, it got worse. We couldn't get anywhere with him, but just the same we never left it. Then I retired as president and he also left the company and another person started.

After Valeria, Doña Rosa was elected. Rosa recounted the difficulties of negotiating with the company, especially at a time when Escondida was seeking to expand their water extraction into the cordillera, directly 'above' the villages of Socaire, Peine and Camar. Escondida presented the project they named 'Pampa Colorada' to the government regulator in 2006.[9] Rosa recounted that by that time she had,

> . . . assumed the role of president with the whole responsibility for restructuring the agreement and just during those dates came this issue of Escondida's petitioning for water up there, above [the Salar and the village], that water in the sector of the lagoons [Miscanti-Miñiques]. So these two problems came together and we were in a tug of war with the company. On one side, we demanded to know why the company had not told us about what they had planned [about new water extraction], and on the other side they wanted to sit down and talk to us about the new agreement, but we paused the thing there, we said '*no po*', we are not going to discuss the issue of the agreement until you stop with the intention to solicit water that belongs here according to ancestral rights and which are subterranean and that obviously will affect the principal flows that we have here in the Community of Peine, and not just Peine, but in all of the communities around ours in the basin of the Salar . . . Well, the truth is that we were a year conversing with the company, between throwing in proposals that they sent, changing them and sending them back with another proposal and so on.

The 'us' of the community during Rosa's time as president included five members of the *directiva* (president, secretary, treasurer and two ordinary directors). Later, it would also include the members of the '*Comisión de*

Tierras y Aguas' (Lands and Waters Committee), a group made up of community members with interests in and knowledge of territorial matters. Nominated rather than elected, the formation of the latter committee, with young and old men and women, expresses a value for Peineño expertise in the increasingly complex matters of territory and extraction. The Committee would provide continuity of knowledge and experience to the annually elected, and thus ever-changing leaders.

In Escondida's Pampa Colorada project, the company proposed water extraction in an important watershed zone for the Salar de Atacama. Community leaders from Peine, Camar and Socaire, and others from the Council of Atacameño Peoples organized protests against the proposal and marched in the streets of San Pedro de Atacama and Antofagasta. People I spoke to recounted with pleasure that when company employees visited their villages, they were refused an audience; 'almost every house flew a black flag on their roof'. In eventual agreement negotiations undertaken in consideration of the threat of water pumping from 'right above the village', as Rosa said, people in Peine became more assertive about the issue of environmental impact:

> Well, the people realized that with the Indigenous Act the waters are subterranean and they are ours by right, and if they take our water they will dry out the river[10] and all the rest. Then we started with the issue of water, but it was never something inside the agreement.

As presidents of the community, Valeria and Rosa represent these events in terms of community members having taken the initiative to make a deal with Escondida, and later having pressured the company to enter a new round of negotiations to improve the agreement. Given that Escondida's license to extract groundwater in Monturaqui would end in the following decade, and the fact that the company had already been sanctioned for generating ecological damage in the Salar de Punta Negra from water extraction there (see Chapter 7), it is likely that the company also sought to use the renegotiation of the agreement as a stabilization of relations with Peine and potentially reduce conflict over the company's Pampa Colorada proposal. After a year of 'arguing', as Valeria says, the parties eventually agreed to a more substantial amount in 'development benefits'.

Not long after they signed the new agreement, the Chilean environmental regulator rejected Escondida's water extraction project Pampa Colorada. Communities around the Salar celebrated the state's decision to stop an unwanted and environmentally questionable water extraction project, and the people of Peine found themselves with a new and improved agreement with Escondida, albeit one that carried with it a raft of technical expectations.

A Plan for Development Benefits

The Development Plan was a technical instrument by which Peineños imagined confronting an increasingly complex future, especially dealing with the problem of extractive industry negotiations and the social and material potential of these, as well as associated community decision making. According to Miguel,

> When we were trying to renegotiate the agreement, the company asked 'so why do you want so much money?' 'How will you administer it?' 'What will you do?' and we thought to ourselves 'we will make *our own* development plan', understand me? So [the development plan] was motivated by the agreement, because if not, they would not give the funds. Because if you say 'give me money to throw in my pocket' no-one is going to give you that money, *po*', but if you say, 'hey would you like to finance a project?' Well, that is another matter.

Miguel represents the Development Plan as the community leaders' tactic for channelling the funds they were seeking to negotiate from the company. For others, the Plan is cast in their memory as something they conceived of locally, even before it was a response to the mining company. Either way, part of the new agreement's terms was that funding must be directed into planned 'projects', including but in no way limited by the completion of the Llullaillaco Stadium.

Using funds provided by the agreement with Escondida, the community leaders engaged known professionals, who then contracted an independent team of researchers, consultants and planners. Across the course of a little over a year (2009–2010), these consultants held meetings and workshops in Peine, field trips with key actors, intensive planning with individuals and community committees, participatory mapping, government document analysis, and meetings with state agencies. Specialists who worked on the Plan included agronomists, geographers, anthropologists, archaeologists and economists. The resulting documents in the Plan are a diagnostic study of the buildings in the village, environment surroundings, infrastructure and energy use and practices of residents, socioeconomic and cultural activities that generate the temporalities and spatial dynamics of place and recommendations laying out projects for the 'development' of each area of life. The Plan was structured into distinct foci: Environment and Agropastoralism; Heritage, Culture, Religion and Beliefs; Tourism; Employment, Organisations and Training (*Capacitaciones*); Education and Health; Sport and Recreation; and 'Other elements of the territory' (including electricity, roads, telecommunications, potable water and water infrastructure).

Research for this book began subsequent to the signing of the second agreement and just following the intensive work undertaken by the community to conceive of and write up their Development Plan with consultants. As such, I relied on people like Miguel, Valeria, Rosa and others to tell me about the recent past, and they sometimes gave different accounts of agreement negotiations and the making of the Development Plan. However, as with Peineños' emphasis in their accounting of the history of Chakizoke, and in Don Lorenzo's characterization of his village (Chapter 1), people cast the process as an example of how Peine community members are strategists of their own future. My interlocutors asserted that the people of Peine community are '*progresistas*' (forward-looking) their development plan is an example of their cultural character.

Don Lorenzo and others of his generation turned around my questions about the social impacts of mining agreements, saying that, as a village, Peine had *looked for* social and economic change, the makings of a good life, and actively sought to bring the world to them. This sense of authority has valence in the way Rosa refers to the Development Plan as a document *the community wrote* in order to secure their share of development benefits from Escondida:

> We were the first community that considered these ideas. You can speak of a city, a country, but a small community planned its own development! We said, 'I want to develop in this way'. Many people from other places have said to me that we were the first community to consider ideas of development, by saying 'I need this, this and this, and I'll capture the resources, and I'll capture them from over there'.

Indeed, other Indigenous communities in the region began to draft their own development plans almost a decade afterward.[11] Such plans are administrative documents and present a series of projects, priorities, costings and designs, and thus seek to manage the community interface with transnational capital and the state. While a discourse of 'development' infused people's expressions of aspirations for improving their quality of life in Peine, the term had specific local meaning and was cemented in a community decision to write a Development Plan as a mode of responding to the 'problem' of community economic responsibility as this was framed by company expectations. I return to this point below, but first expand further on how the content and function of Peine's Development Plan troubles the value of 'development' as a universal modernity and as securing monetary benefit, and articulates the moral economy of agreement in Peine. The Plan itself and the memories of those involved in creating it run parallel to, as well as exceed the agreement with Escondida in important ways, especially in the sense that Peineños speak about the Plan as representing a future relation with their waters, a future conceived in their own terms.

Development as Future Relations

There are half a dozen large files and bound volumes of reports and documents that make up the material record of Peine's Development Plan.[12] Much of the documentation is dedicated to matters relating to 'environment': natural resources, surroundings, their uses and an examination of the relation between economic and ecological capacities and community subsistence and flourishing. The Plan is rich in detail. Each section includes data collected from government reports, data generated by describing Peine's local conditions, and accounts from group discussions with Peineños. It is designed to produce an analysis of every conceivable aspect of social and infrastructural life. In the final part of the Plan, the consultant authors expand on two kinds of conclusions: first, solutions proposed by community members who participated in workshops, and second 'the consultant's vision', or how solutions might be addressed in technical terms. These are then elaborated in outlines for future projects and priorities. As a Plan that community members involved spoke to me of as belonging to them, originating in their own ideas, it is also a record of translation between Peineño visions of community futures and technocratic efforts at fungible designs. However, some of the projects in Peine's Plan, especially their vision for environmental flourishing, confronted Escondida's vision of what counted as 'development'.

In the diagnostic section for 'Environment', consultants identify four key environmental 'problems'. As with each section of the report and plan, there is an account of how community members workshopped the diagnostic material with consultants. For example, in a mix of third person and passive voice, the consultants report the following environmental problems:

> There is no capacity in the community to manage the natural resources.
>
> There is no monitoring or control on the Community's part of the environmental promises of projects executed in the territory.
>
> They do not take advantage of renewable energy available.
>
> There is a perception that biodiversity is being lost in terms of both abundance (the number of individuals of each species) and richness (the variety of species inhabiting a particular area either seasonally or permanently).

(Plan de Desarrollo, property of the community, page 4, repeated in the report on page 37).

The community plans to deal with these problems by increasing their 'internal capacity'; training programs for community members, creating professional networks with universities and public institutions that can assist them, and forming a stable committee of persons responsible for 'the environment' made up of Peineño experts and professionals. The latter resulted in the formation of the Comisión de Tierras y Aguas, mentioned above. The second of the four points refers *indirectly* to the experience of community disempowerment in the face of environmental impact from the mining industry, and the passive voice is used to refer to the 'perception' of biodiversity loss. Thus, while community members identified the 'environment' as needing the closest attention in terms of the future of the Community, the consultants who wrote the Plan document were careful, and in some places euphemistic about the expression of these concerns.

External pressures on consultants were manifest in a meeting I attended during fieldwork in late 2011, when they reported to community members on preliminary conclusions of one of the priority projects: an environmental monitoring study using remote sensing data and focussed on the wetland areas and ecological zones on the southern end of the Salar and adjacent steppes.[13] The geographical area of their studies took in key sectors of Peine's territory, both lowland and highland water sources and pasture. Consultants reported to the community that when they had presented the preliminary results of their environmental impact diagnosis of the area to Escondida, company employees instructed them to cease working on the material because Escondida was undertaking environmental impact studies throughout the area. As my fieldnotes reflect, the consultant informed the meeting in Peine that they had been asked rhetorically by a company employee, 'why should we pay for this work twice?' As community members in the meeting were quick to see, Escondida was not prepared for them to use development benefits from the 2007 agreement to engage independent environmental consultants to undertake environmental impact studies relating to Escondida's extractive impacts in the southern Salar.

There was considerable community anger and disappointment expressed at this project stymied. Miguel noted archly that during Escondida's recent visit, the company's environmental scientist said that his recent analysis showed that the waters in the southern end of the Salar had actually risen not dropped. I was also present at that meeting and had heard the company scientist report that data drawn from their monitoring bores showed water level rise and fall in more reliable ways than remote sensing data.

Consultants' preliminary remote sensing studies, as reported in the Development Plan, point to environmental harm in Tilopozo as a result of industrial groundwater extraction. In the meeting with the consultants, Jonatan, who had been active in the workshops with consultants, asked the consultant rhetorically, 'why would the company pay for you to do the work and then speak about your work in this way?' He continued ironically, 'Since we have signed an agreement with the company, are we not *socios* (partners) in such matters of development with the company?'. His sarcasm belies the ongoing distrust of Escondida, despite Escondida's assertions of being 'partners' with the community for development and its claims in company environmental reporting. An ex-president of the community noted urgently that they must act on the consultant's environment report and act on it well, because 'the eyes of the region are on us'. He continued, 'Many people know that Peine has this agreement, so they can fund their Development Plan and people are watching what will happen . . .' And he cautioned that if they take Escondida's money and sat back for years until the agreement was finished, so then their water would be finished, and, he asked, 'what will we do then?' Such concerns parallel the Development Plan document in which the consultant presents their 'vision', as based on year-long discussions with community members;

> One of the principal riches of the community is its territory and the resources found there. The current generation receives what their ancestors knew regarding how to conserve and be cautious. It is the local people that feel ethically obligated to pass on to the young generations and those still to come, a habitable place equipped with its inseparable resources such as the beauty of the landscape and the availability of resources, in a way to ensure biological and also cultural survival. From this point of view, the Development Plan should contain measures that permit the Community to achieve a growth in their quality of life, including the environmental aspects, even when mining activity (that today provides employment and support to the community) reaches its end.

The vision counters the value of development as economic gain through unimpeded extractivism by identifying the need for studies of their environmental surroundings independent of company data, and rhetorically distances the community from the relationship with Escondida by projecting an emplaced future among their neighbours at an imagined horizon where mining was finished. Continuing to 'manage' the waters most impacted by mining is central to how people in Peine describe what the value of development entails. This management extends to how their actions are perceived by neighbouring communities, with whom they share relations to water and territory.

Peine's Development Plan is in part a local strategy of responding to the mining industry by which the community has articulated their own logic of development, often in distinction from that espoused by Escondida. This was complicated by the fact that the company retained control over how funding could be used.

In order to execute an individual project from the Plan, community leaders must demonstrate that there is consensus in the *asamblea* for a project to go ahead, that they have adhered to technical details by engaging professionals to quote on the work, and that they then conform to strict financial auditing procedures. Proposed projects have languished because they do not meet these conditions or, as in the example given above, may be rejected by the company.

In 2011, the benefits from the development fund were piling up in a trust account that held them, and spending overseen by the *Comité de Adelantado* was not occurring. I was invited to a meeting of the committee by the newly elected president Don Guillen, the permanent committee member from the community, Don Lando, and Director of FME. Functionaries from Escondida's social relations team were absent on this occasion due to car trouble, we were told.[14] Items on the agenda included squaring up the paperwork for Peine's leadership changeover and a discussion about unexecuted development fund projects. The FME Director read officiously and laboriously from the accounts in front of him, then removing his reading glasses and holding them at an angle, reproached the two local men, saying, 'You really have to spend this money, get the projects going that were recommended by the university'.[15] He complained that no funds had been spent *properly* since 2008. Neither of the Peine older men in the meeting mentioned Escondida's denial of funding for the community's bid for an independent environmental impact study.

A document like Peine's Development Plan instantiates aspects of progress discourses embedded in the moral economy of CSR (Rajak 2011a); if the Development Plan enabled development projects, then the community-company agreement could be said to be operating well, and extraction as development a social value, morally justified. I mentioned to an Escondida social relations employee, when they visited the community a week after the meeting described above, that I had observed the FME Director and his push to spend the funding. They complained that like him, they too were under pressure from *their* bosses to get Peine to spend their benefits and show some development results. People in Peine I spoke to during this time were also concerned that they had an ambitious Plan, *muy complejo* ('really complex'), and few finished projects. Thus, a company-community agreement such as Peine's with Escondida could be read as initiating a relationship with mutual obligations to see development outcomes. However, as the events I describe demonstrate, in no sense were the parties to agreement *in* agreement about what those obligations entailed, or how the relation would function. Instead, what is most prominent is evidence showing that mining *as* development seemed to deliver few tangible benefits despite a few village infrastructure modifications (that Don Bartolo thought might only 'make things *pintoresco*'). The relation has functioned well in delivering contested visions of environmental concern.

Escondida refused the idea that its development benefits funding might be used by Peine to fund scientific experts independent of the company. Peine's elected leaders, in their own voices and those of their hired professionals, have continued to insist that independent environmental studies are necessary. Sometimes these assertions get caught up in questions of 'trust' and company transparency, framed by a politics of knowledge (see Babidge 2015). However, given the apparent enormity yet opacity of Escondida's operations, and the ways in which the company has actively dissuaded the community from independent studies of company environmental impact, the parties to the agreement are so at odds, their interests may be incommensurable. What is clear is that contractual agreement for 'development benefits' neither mutes the community's ecological concerns nor their assertions that these concerns trump all others.

* * *

As narrated in Chapter 1, at the reinauguration of the Lllullaillaco Stadium in Peine, community members, employees of Escondida, FME, and local government representatives gathered to celebrate the finalization of twenty years (in 2017) of interrupted construction. My own attendance at the stadium inauguration was unplanned, a stroke of anthropologist's luck.[16] Perhaps partly due to my sense of 'being in the right place, right time', the event seemed an anti-manifestation of environmental concerns that dominated engagements I had witnessed on other occasions at first glance; it seemed a co-constituted suspension of company-community contested relations-as-normal. However, the stadium reinauguration event did not resolve any other process except the unfinished stadium, nor was it a celebration of a 'good' relationship as the FME Director would like to believe, or other company actors expressed during their own speeches. Instead, it coexists with the disjunctive relations which have continued over time. The parties to an agreement sometimes hold open contestation in abeyance to enjoy a rare moment of extractivist development benefits.

As for many different jurisdictions and as has been demonstrated in studies by legal scholars and social scientists, Chilean laws relating to Indigenous rights to lands and waters do, in theory, recognize Atacameño peoples as the owners of their lands, but the terms of these territorial rights have not seen full recognition of claims by Atacameño peoples (Yáñez and Molina 2008). In addition, all four of the Salar's principal mining operations were installed and their water extraction concessions granted prior to any such recognition measures. As such, negotiations between Escondida and Peine have been characterized as 'voluntary' by companies in a way that parallels Rajak's (2011a: 18) analysis of CSR as a powerful gift in a universalizing moral economy of 'ethical capitalism'. Transnational extraction companies' concerns about their international reputation (increasingly relevant to investors) and national reputation in terms of gaining environmental permits has relied on such claims to be doing 'good business'. Escondida's early agreement with Peine is an example of the way in which that company, owned by two of the major players in the global politics of corporate social responsibility, audit and transparency – BHP and Rio Tinto – has been part of a corporate movement to trumpet standards of social responsibility in spite of inexistent or ineffectual state regulation regarding Indigenous rights and environmental protection, and despite the fact that there has been ongoing community dissatisfaction with the company's activities and indeed the agreement itself.

The idea that the company has been 'better [than the government, than other companies] at recognition of Indigenous rights', as claimed by one Escondida manager I interviewed, ignores a long history of how, remote from view, the company has sought to avoid matters in the material demands of the Peine community. Escondida employee, Nicolás told me in 2013,

> I think that we have to continue with the agreement. What we are clear about is that we cannot leave the community alone. That we are there already and so we have to continue and that is part of our values as well, to maintain the relation.[17]

However, discussing the environmental and political fact of their relationship in 2016 with a then recent replacement of Nicolás at the Escondida/BHP offices in Antofagasta, he told me, 'we would prefer not to have to have a relationship with Peine'. He predicted a time when their desalination plant would, he proposed, allow Escondida to effectively remove their water extraction from Peine's territory, and imagined that this would end the community-company relationship. The future relationship between the company and the community would be characterized by an abstract past. Notwithstanding cessation of Escondida's groundwater extraction, the lingering questions about ongoing 'impact' from those activities remains.

Conclusion

The Peine-Escondida agreements, as legal instruments, did not settle the terms of relations between the parties, meaning the company continued to deal with operational uncertainty, nor did they satisfy the political, economic and ecological interests of the community. The need for a relation between the company and community is premised on the water 'sucked out' of the southern Salar by the company, and the community asserts that this is doing ecological harm, and that it will continue to do so in the future – as Miguel said, 'produces, that *is going to* produce' – especially in the place called Tilopozo. The agreement does not address these ecological concerns. Nonetheless, territory and water relations give rise to and nourish the Peine-Escondida agreement and its conflicts. 'All the subterranean freshwater of the southern Salar de Atacama subject to extraction by Escondida' may be one way to describe the body of water that mediates and also divides all community-company relations and that sits beneath the surface of the agreements they have made. The absence of water and environment in the agreement's terms and yet groundwater's centrality to the relation between the parties is an awkward tension at public occasions and meetings of the parties. While muted by technicalities, the ecological effect of extraction to groundwater refuses to be left out. As a subject it courses through community-company business, its insistence sustains social tension between the parties at contract.

Concerns for water and territory may impel community decisions to sign agreements with mining companies in the first place. Ecological concerns may have been strategically muted by the community in coming to a contractual agreement, but the lack of environmental provision in the Peine-Escondida agreements has resulted in structural and legal limitations on community in favour of emphasis on the effectiveness of agreements for community economic benefits. Nonetheless, 'benefits' may be difficult for the company to transfer in ways they wish, because there is often no agreement between the parties about what constitutes a benefit and how economic funds are to be spent. Buildings and infrastructure have been useful, as the stadium stands, a tangible outcome from the relationship. They are also a structural reminder of the problems. To return to the Llullaillaco stadium, at the time it was finished, Escondida lodged an Environmental Impact Assessment with the Chilean Environmental

Evaluation Service (SEA) in which they sought to renew their concession for water extraction from the southern Salar de Atacama after their license was set to expire at the end of 2019. As part of the EIA, Escondida offered a range of corporate funding for infrastructure, education and economic development projects (which they represented as 'voluntary measures') to Indigenous communities of the Salar de Atacama, including the Peine Indigenous Community. The community continued to insist that Escondida must cease all water extraction in the southern Salar. Peine thus rejected the idea that economic transfers from the company through their existing agreement or any other activity would oblige them to agree to turn away from Escondida's environmental harm.

Notes

1. This is crude language and conveys the tone of Miguel's speech.
2. Following ratification of C169, Chile introduced regulatory mechanisms to institutionalize and proscribe Indigenous participation in extractive governance. While laws were introduced in 2010, these did not begin to take effect in the north until some years later (see Chapter 7).
3. For example, in 2012 the community leaders asked me to make enquiries for them in Santiago about legal representation and professional support from environmental and Indigenous rights NGOs. The answer I received from two prominent NGOs was that on principle they only work with communities that resist mining and not those who sign agreements.
4. As outlined in Chapter 3.
5. The establishment of such boundaries is not free from inter-community conflicts, especially in relation to the ownership of vegas (pasture).
6. For example, during a community field trip to Monturaqui as part of the consideration of Escondida's 2017 EIA, older people directing the bus became confused about which road we must take because of all the snaking diversions.
7. Don Geraldo did not travel to the mine site with the others, Valeria said, because he did not pass the health test. When I asked him about this he was suspicious at the result. How could he not have passed their health test to enter the mine, given his life history in the *puna*, daily labour in his fields and general good health?
8. M. Ojeda was the Director of FME for a number of years in the early 2000s. He published a piece about sustainable relations with communities (Ojeda 2002).
9. The case of resistance by Atacameño communities to Escondida/BHP's proposal to extract water from 'Pampa Colorada' is detailed in the account by anthropologist Paola Bolados (2014), relying on extensive interviews with community leaders across the region.
10. Some people (such as Rosa) refer to the flow of water from the principal springs of the village as '*el rio*' or the river.
11. These have a new language, as Planes de Vida or 'Life Plans', in parallel with Indigenous politics across the continent, emphasizing post-development philosophies and the '*Buen Vivir*' or good life (see Blaser et al. 2004, Escobar 2012).
12. The Plan is not a published document; it belongs to the community. My description of its contents relies on the notes I took in 2011, during a week's detailed study of the

documents. I seek here to describe content relevant to the argument at hand without revealing unnecessary detail (as agreed with community leaders as a condition of access).

13. The area of the study was a transect of Tilopozo and the salt lakes on the southern Salar, and from there east through Tilomonte's main canyon toward the Pular peak, a project named 'Action Plan of Conservation of the wetland pastures in the Tulan-Tarajner, Laguna Salada-Saladita and Pular transect'. Remote sensing data uses a series of satellite imagery to analyse landcover change, earth surface temperature, and other factors. It is especially useful in analysing environmental change, where access to study of the environment is limited by remoteness of geographical area and commercial ownership of environmental impact data (Babidge et al. 2019).
14. While the agreement included a role for CONADI, I do not know what this role entailed, and I did not witness or hear about staff from this state agency participating in relevant company-community meetings.
15. The consultant group that worked with Peine on its development plan included scientists and social scientists from the University of Chile.
16. I was lucky like this a few times with different events, that is; I happened to arrive in Peine to begin a period of fieldwork on a day or during a week that for example, a company-community meeting was being held or some other event was occurring, and in which I had interest in terms of this research. Leaders of the community seldom informed me of such matters ahead of time, nor would I necessarily inquire about such events in communication with leaders prior to my arrival. Meetings with mining companies and others were an increasingly regular and increasingly burdensome matter (see Chapter 7), thus my research 'luck' was less and less surprising.
17. Direct quote: '*Yo creo que hay que seguir con el convenio. Lo que si tenemos claro es que no podemos dejar a la comunidad sola. Y que una vez que estamos ahí, tenemos que seguir y es parte de nuestros valores también, mantener la relación*'.

Chapter 5

GOOD WORK AND 'SHARED BENEFITS'

General Pinochet came to Peine in the 1980s. The visit appears in a propaganda magazine produced by the regime (Figure 5.1). There is Pinochet, infamous architect of violent oppression, smilingly greeting a group of women on the streets of the village. I showed the image to Doña Valeria, who remembered that she and others had been asked to prepare food that day in readiness for his scheduled visit to the Mother's Centre.[1] She described it by saying that,

> Pinochet gave a speech promising many things, but these things never arrived. Then he got in his helicopter, which was parked up on the hill . . . and went to the lithium mine. My son insisted on going in the helicopter with them and they dropped him off in the gardens before Pinochet flew over the Salar to inaugurate *El Litio*, the plant and mine.

Valeria finished with the rhetorical question, 'Perhaps this [the inauguration of the lithium plant] is why he was here?' and answered herself, 'must have been'.

While the state has been absent for much of the history of the region in the sense of providing for remote village citizens, it has in many ways been present in the impulse for, and sometimes the ceremonies associated with, the expanding frontiers of extractive capitalism. During the early years of the military dictatorship, Pinochet's civic-military administration provided for the Salar's development in concessions granted to lithium miners. In an extractive industry dominated by transnational interests, the Chilean state kept a material and symbolic relationship with the Salar. Pinochet's regime pronounced lithium a 'strategic mineral' by legal decree in 1979 (DL 2.886), given its potential in nuclear fusion (Gundermann and Göbel 2018; Barandiarán 2019b). CORFO signed an agreement with the north American company, Foote Minerals in 1980, for Foote to extract

---

Notes for this chapter begin on page 153.

Figure 5.1. General Augusto Pinochet visiting Peine c.1984 (copy of pamphlet, from private collection held by author).

lithium under the company name 'Sociedad Chilena del Litio' (rendered SOCHILI or SCL), providing a concession of over 160 square kilometres of the southern Salar to the extractive operation, and later gave a concession to SQM, to the north.[2] SCL was a transnational operation with an important symbolic link to national belonging.

Life in Peine prior to the arrival of industry was '*puro sacrificio*', some people told me, Don Geraldo emphasizing '*la vida era mas jodida, mas crítica*' (life was difficult, precarious).[3] Relying on agriculture and pastoralism was, as this metaphor goes, a 'sacrifice'. Peine before the lithium mines was a poor place with salty water and few resources, people said, comparing it to the neighbouring village of Socaire, where sweeter and more abundant fresh water at higher altitudes feeds pasture and crops in greater volume and variety. Older people described a salt-infused sense of abounding scarcity and a readiness to respond to new and promising opportunities. The period after would be characterized by paid labour, the possibility of opening small businesses and new possibilities with changing corporate relations. Comparing life before in relation to contemporary material comfort, small scale agriculture as primary livelihood appears in people's memories as an act of surrender (either of oneself or another) to the eternal round of putting the body to work; labour to sustain living. The language of sacrifice describes how livelihoods are remembered as conditioned and conducted by the limits of salty ecologies, but the memories of constrained and marginal existence were also *tranquilo*, relaxed and quiet. When El Litio began, many older Peineños said, the calm but also materially sparse life changed.

The location of the village led to early and intimate labour relations to the brine-extracting lithium industry that started slowly in the Salar, first with exploration in the 1970s, the opening of the SCL mine in the following decade, and the escalation of change brought about by the lithium boom from 2012 onward. SCL began extractive operations in 1984, and remained for over two decades a small operation built on the foundations of Pinochet's military state strategies and a shared enthusiasm for the operation from locals who enjoyed employment and continuities of life in Peine that the work nearby enabled. SCL was an operation with fewer than 100 employees, a majority of whom had a permanent address in Peine, and local managers and bosses who were Chileans.

In the first years of my fieldwork in Peine, almost everyone I spoke to referred to the mine and the company simply as 'El Litio' ("The Lithium[Mine]"), even though 'Minsal' (now SQM Salar), a much larger extractive operation, broke ground further north on the Salar soon after it. In 2012, 'El Litio' began trading under another (majority) north American company, Rockwood Lithium.[4] In the changeover, the community and company negotiated a formal agreement, which set out the terms of 'shared benefits', by which the community would gain a profit share and the company would proceed with expansion of the lithium mine (see chapter 6). During a period of fieldwork in 2012, when asking people in Peine about 'El Litio', they first corrected me with one specific company name (Rockwood), and later, during a period of fieldwork in 2015, the other (Albemarle). These are awkward names to pronounce in Spanish, and the linguistic incongruity made me consider the ways that 'El Litio', long a familiar place of work that relied on local labour, was transforming. The referential shift marks a formalization of the relations between company and community; having previously operated according to the informal intimacies of patronage, they were now parties to an agreement. It also marks a period in which both lithium mines on the Salar were expanding, and the area was shifting into focus on the global and national extractivist stage.

Fluctuating material politics of the lithium industry, especially as these arose in conversations with older community members, are the focus of this chapter.[5] Life history conversations with around twenty-five middle aged and older people were recorded during fieldwork between 2011 and 2017, with some intensive interviewing of half this number undertaken in collaboration with a local research assistant ('Amelia') across six weeks in 2013.[6] In the stories that older people told us, life changed radically in the second half of the twentieth century. In fact, 'change' is what seems to have marked life itself from the 1970s onward, with the 'before' in recounted memories pulsing with a sense of rhythmic constancy. Ecological, social, and political forces of various kinds intersect, and together inform a sense of the changing present. In the telling, people emphasized material – economic and ecological – constraint and potential, the evaluative explanation of how and why things need to happen in the ways they do and opinions on how they should. The contours of these stories are particularly important because of the period in which we recorded the oral testimonies, which was exactly as speculation about a lithium extraction boom was emerging.

I begin the chapter by recounting how people in Peine narrated to us their labour with saltwater and with the salt lakes, and their participation in the initial stages of mineral exploration on the Salar itself. Diverse perspectives on how to make a good life within the ecological constraints of the desert arise out of such memories. Peineños' relations with 'El Litio' saw these relations both challenged and enhanced. Exploration in the southern Salar found that the lithium-rich brines were very promising. For locals, who enthusiastically sought work in the mine, the demands of their fields and families were combined with an industrializing Salar. The material relations between El Litio and Peine, their intimacies and informality, and the ways in which these relations shifted considerably with the negotiation of a formal agreement in 2012, are narrated in the final section of the chapter. The long-serving manager of the plant is the unavoidable symbol of company change. As global lithium speculation heightened, along with the plant's expansion, lithium mining appears increasingly more like the pattern of dispossession stereotypical of the industry in other places. Nonetheless, it carries with it the local history of intimate and subtle negotiations of shared experience and identity.

Before 'El Litio'

The term 'salt pan' conjures the image of an expanse of hard, dry earth, but the Salar is a varied place, with its surface inhospitable to easy travel and marshy areas in which animals and vehicles may become stuck. Don Geraldo emphasized that one needs to know the Salar 'to walk well' on it. Peineños used specific paths for generations to reach areas of seasonal ecological abundance, such as lakes and raised salt crust. Large blocks of salt were cut from a specific location. These would be strapped to a mule's flanks and transported for local use and trade to neighbouring villages, and beyond (see Molina 2017). With some levity, I was told that the old nickname for Peineños, '*lomo salado*' (salty loin), is a play on their salt trade, the proximity of the village to the salt flat and its salt-tinged waters, crops and bodies. Salt was traded with the '*bolsas suicias*' (dirty bags) of Toconao – the double entendre referring to testicles and the hessian bags stained from carrying tree-fruit crops – and with the '*patas rascadas*' (cracked hoofs) from Socaire, where use of *ojotas*, Andean sandals in the higher altitude and harsher climate hardened the skin of their feet. As pastoralists and irrigators of maize, grain, alfalfa, and potato crops, these villages differ in the size of their herds, productive potential of harvests, diversity of crops and the quality of their waters. The nicknames are signs of older material interrelations among the Atacameño-Likanantay peoples of the Salar and their own bodily and practical differences as these belong to each oasis and niche ecology.

Peineños know the Salar, not just from pasturing animals in the southern end where freshwaters and marginal grasslands appear at Tilopozo and Tilocalar, but also the dry areas where block salt may be harvested, and the interior lakes where Andean flamingos (locally *parinas*, *Phoenicoparrus andinus*) breed. Until around the early 1970s, people from Peine entered the lakes at the end of the first week of December for the annual harvest of *parina* eggs. Around February, they would enter again adorning hatched and growing chicks with decorations in the same way that herds are subject to *floreamiento* (lit. flowering, see Kalazich 2013), and performing a fertility ritual.[7] Rosa (in her early 40s) said,

> My mother told me that in the old days, when she [the mother] was young, they would go down where the *parinas* were in Salada and Saladita Lagoons and they would sing to them, dance with them and dress the *parina* chicks like babies. They would do their ceremonies for a time and then let them loose, doing no damage to the birds, and only taking certain marked eggs.

Valeria recounted that for the *parina* egg harvest in which she participated during the early 1970s,

> Everyone entered [the Salar] . . . there were not so many people doing this as before, but there were still people there and sharing out the eggs among families. The numbers of animals were huge. They would fly over in the night and then when the chicks had grown, they would fly back to the cordillera again.

Feathers collected from the *parinas* are saved for the *limpia de canales* and gifted to the spring waters, linking all the beings of the territory through agricultural labour, the peaks, the village and the Salar.

In 2013, when we met to talk about his working life, Don Ángel had just the year before retired from El Litio. He suggested we meet near his fields and we perched on the stone wall, the digital recorder between us. He began our conversation recounting how the people of Peine built the wall to divide the gardens and fields from the houses. Each gate has a name corresponding to the sector of fields into which they open. When Ángel was a young boy, in winter, when the post-harvest husks of corn were dry and the weeds and alfalfa shoots sprouted amongst them, people released their sheep and donkeys into fields to graze and leave manure to fertilize the following years' crops. After fields were sown, a person would be fined or punished for allowing animals through the gates. Today the wall does not serve this same purpose, the gates remain open, and chunks of the wall have crumbled. Small herds of sheep and goats are kept in corrals adjacent to fields and fed from alfalfa bundles grown and dried especially to sustain them.

Ángel was a teenager when he and his brothers left Peine. He told me that foremost in their minds was the desire for nice clothes and shoes that one might choose for oneself, not pants that showed the place where mother had patched the seat, or the *ojotas* (sandals) they had worn to school. At the age of 15, maybe 16, Angel went to the city:

> I was working in Antofagasta first. Clearing plates in the restaurants there, clearing plates, cleaning glasses, cleaning. I was working there for little more than a year. From there I went to Socompa. Back then everyone went on foot, *po'*. On foot! Walking with the animals. Before now, everyone walked everywhere, *po'*. There are the tracks there where the animals passed through, around one hundred or two hundred animals. From there I worked in the lands around Socompa, unloading the bulls once they arrived [from Salta on the train]. In winter it was damn cold that business! And sometimes the train arrived at 3 in the morning, sometimes 5 in the morning, and you just have to get up and unload them. The cold! Yes, then in the afternoon after having fed them, given them pasture and water . . . we'd move them over to the Chilean side and there load them onto the train again, *po'* . . . a Chilean train . . . sure, life was pretty sad in those years. After that, I did military service in Antofagasta, then I went to Ollagüe, near the border with Bolivia, and worked there for a long while, and then went to Calama to work . . .

The railway, which began construction in the 1920s, stopped at Socompa (on the Argentina side) and began again at Monturaqui (on the Chilean side). Men and boys from Peine worked on both sides, moving herds of cattle and cargo from one end of the line to the other. Monturaqui station was a small town; women worked as cooks, some leaving children in Peine with older family members so they might attend school.[8] Don Bartolo told me, 'And we were almost all there. Yes, we had a football team there, all men from here'.

Before El Litio, as Doña Lucia said, 'there was no electricity, none of the comforts of today with the water, we didn't have tele[vision], so we went to bed early, *no teníamos na' pa' salir* [we had nothing to go out for at night]'. As she reminisced,

> The school teacher [c.1960] told us that mining would come and it would change the pueblo. We never imagined that it would happen. 'You're going to have work', the teacher told us. El Litio arrived slowly, CORFO, a few trucks. Then there began to be things for sale [in locally run shops]. Before, we made bread every week in the wood fire here, now we buy it . . .Thanks to El Litio we have so much, El Litio did a lot. Before we got around on donkeys. *La vida muy sacrificada* [Life was a sacrifice].

Doña Lucia stayed in Peine to help her mother, but her older sister left when she was very young, working in a mining city 'for many years. She did housework, . . . in the house of [a Yugoslav] family. The señora had two children. My sister grew them up, [as] they say, grew them up'.[9]

Many left the village when still teenagers to find work in other towns and cities of Chile's north, most connected to the mining industry. Those who remained to cultivate fields and look after the herds scraped a living. As Paulino, a man in his late 60s, who migrated to the south for work as a teenager, told me, goods like 'sugar, flour, oil, soap, all those things. They brought them across and sold them here [in Peine], we practically got fed from Argentina'. If they sought commercial merchandise such as noodles, rice, clothing, tools and other materials that demanded cash, they might sell an animal, taking sheep to San Pedro de Atacama to sell, or perhaps '*el otro lado*', the other side of the cordillera, Argentina. Claudia, born in the neighbouring village, Socaire, and who married a Peineño said, 'When I was a girl, I knew Peine as a poor place, . . . in the 1970s it was still very poor. Only in the 1980s *levantaban gente* [people came up]'. It was then that she moved to Peine for the opportunities that arrived with El Litio.

With the *Golpe*, the fear of state violence arose to threaten local life, reaching into the southern Salar, and with National borders closed, the railway and its work for the people of Peine also finished.[10] Perhaps a decade prior to the Dictator's helicopter landing, Peine and neighbouring villages were occupied by the regime's military personnel and the territory was planted with fields of land mines, as Valeria recounted:

> Around '73 at the beginning of the *Golpe*, the military was here in Peine, *cualquier cantidad de militares* [huge numbers of military men]. They had explosives

> down in the old camp. I went to Socaire to spend summer holidays with my mother and there were *militares* there as well. They made a curfew and sometimes the militares would go house to house looking for people. In Peine [three named men] were taken from their homes and accused: *pescaron, amararon y declaron que eran comunistas* [they found them, tied them up and declared them to be communists]. I don't know who accused them . . . People had lots of guns in those days, for hunting or protection in remote parts of the mountains and were supposed to declare them. [Valeria's father] only declared his pistol, which they let him keep. The rest he hid in the gardens and some of them my brother still has today. Others did the same.[11]

One regiment was camped at the edge of the village near the *playa*, where the irrigation canals empty out onto the deep sand on the edge of the Salar, and another was camped in Tilomonte, where soldiers camped in people's gardens, trampled the crops and stole their corn. Elected members of the Junta de Vecinos (JV) were replaced by people whose pasts were not marked by union membership or leftist political histories. Military posts on the border, fields of landmines and confiscation of weapons prevented men and women from travelling with their herds into and across the mountains in summer.

Throughout the early 1970s, CORFO undertook exploration for minerals and oil in the Salar, at a time when older modes of life for Peineños were quite literally drying up in a widespread and long-lasting drought that radically reduced pasture in the broader territory, as well as the reduction in possibilities for work due to military occupation. Herders like Doña Inés continued to take her flock to Tilopozo and saw the state and mining prospectors begin to appear. She narrated this seemingly preserved in the sweet memory of the gift of tinned fruit.

> CORFO was around. I think they were, what do you call it? *Haciendo mediciones* [Doing tests]. They were there exploring, I think, since they had trucks and they built a shed for their equipment. And as the children were always running around, the men stopped the truck one day and said, 'hello children,' [laughs] gifting them biscuits, tins of apricots, and juice.

While Inés and others continued the older form of life, many men from Peine were employed by CORFO to assist in the construction of roads and exploration bores for future lithium extraction.

'Estaba Buena la Sal!': Working for a Good Life

'The salt was good!' Don Bartolo said, with brines thick and viscous. Bartolo and Benicio were young men who gained work with CORFO in exploration and construction. They and others gathered samples, cleared ground, and excavated experimental evaporation ponds for lithium production. A road to Tilopozo is marked by a milestone that reads 'zero kilometres', which implies that nothing was there before and no other company had begun operating in the territory. Don Bartolo was with CORFO when it began. He recounted,

> We went looking, every 500 meters making a drill hole. In the first 500 meters we found stone, just by drilling a little. The next 500 meters, another and the rock was a little deeper, as it is like a horn [indicates the shape], like a cone. Rock, rock, rock, until we got to Kilometre 20, as they call it there, and we were drilling about 300 meters below and, we couldn't keep drilling because the rotation drill kept getting stuck . . . It would come like this [makes slurping sound] . . .

Bartolo and others were involved in surveying, construction, measurements and eventually in the lithium plant and laboratory. In the late 1970s, non-local contract workers were camped near Peine's swimming pool. The government department of land and property, *Bienes Nacionales,* then excised an area at the bottom edge of the village (where the camp still stands today) for offices and a dormitory. There were no forms of community consultation regarding the impacts of development on villager's lives, and no recognition of people's territorial authority. Bartolo remembers the trouble the workers caused sometimes in the early years, but the small scale of the mine and the form of labour it offered to local men meant he and others remember its beginnings with a sense of the busyness of the work and only occasional rancour at the intrusion.

Many young people returned from elsewhere, looking for work, both men and women. Claudia, in her late 60s when we spoke, met a Peineño in Calama and came to work in El Litio as a cook. Some men, '*toda via eran cabros*', were still very young and excitedly sought work on the Salar. The 'Chilean' business and its long-term contractors employed men and women from Peine and those who came from other Atacameño communities (especially Socaire), as well as Chilean men from outside the region.[12] Women and men took up opportunities to earn cash from the burgeoning industry, starting businesses to cook meals, bake bread, and wash clothes. Two people opened *almacenes,* corner shops that stocked tinned and dry goods, cigarettes and snacks to sell to the contractors and employees.

Don Ángel returned to Peine to help his mother prepare to plant their annual crop of corn, and upon hearing that the lithium mine had started, he enrolled his name at the company office. In 2013, after more than two decades employed in El Litio, Don Angel was a man of some means. He drove a late model *camionetta* (pick-up utility) and could often be seen dressed in a button-down shirt tucked into pressed pants and clean boots (when not in the ubiquitous coveralls leftover from the lithium plant worn to work in the fields). Ángel's account of first gaining work reveals how personal the operations were at the beginning. After he had put his name down in the company office, the lithium plant manager came to see him at his mother's house.

> This *caballero* [gentleman] asked me, 'do you want to work?' '*ya po*', I told him, and I went to work. As it happened, my mother was in Qene [agricultural sector] with the herd, and I took the donkeys down there to pasture. I said to my mother, guess what? Don Tomás asked me to go work and I want to work. She said, *bueno*. It was a little before September 18th.[13] And I said, okay, but . . . let's plant the fields and then I can start work and not have to ask for days off. '*Si, po*', she said. *Pucha,*[14] right away we had to do our planting, *al tiro*! I asked Don Tomás if he had a place for me after we finished planting. This gentleman came to see me again and said, '*ya po*', he said, 'once you're not occupied, *vai pa' adrentro*', [go into the Salar].[15] So, we planted, left everything planted, and then I went to work, and I stayed working there for 24 years. Eleven days on, three days off.

Working inside the Salar and lithium plant demanded temporal accommodation in the cycles of agricultural practice, and a shift in pace at which tasks were accomplished. Men and women talked about mining work in El Litio scheduled as twenty days on, three days off, later eleven and three, and most recently seven by seven, with twelve-hour shifts. Men like Ángel tell stories of El Litio with enthusiasm and ease; the planting and harvesting seasons were more pressured, but could be fit into the rest days, in addition to lighter agricultural tasks and tending to animals at the beginning and end of a day's work. Don Benicio was also one of the first to work in the lithium mine and recounted that during *descanso* (off-shift, or break), 'in that time we planted, the changeover was quick'. Interviewing with me, Amelia, Benicio's neighbour asked him, '*Vecino*, it wasn't an impediment? Having to leave aside agriculture?'. He retorted,

> Maybe for other people. I just continued. I always had to . . . I worked all my life, *po*'. Now, sure, I go with the Seniors Club on travels, but I don't have much money. When I had my herd of sheep [that I had before] you'd have to go in the morning and the evening [to feed and water them].

Benicio's proud answer gave a sense of how some were able to keep up this dual temporal and labour responsibility. Others began to reduce their herds. Don Bartolo lamented how waged labour gradually saw the end of the practice of *minga,* of working together among neighbours. He said; 'now one doesn't turn up *con el azadoncito* [with your 'little' hoe] and say, *"yaaa vengo a minguear"*, that, one doesn't see anymore'; there is little of this kind of work as solidarity and for the feast that follows, rather than for payment.

At the time Amelia and I were undertaking oral history interviews together, a documentary aired on a popular television channel that showed an Atacameño village populated by only a handful of aged permanent residents. A few people we spoke to commented on it, saying that if mining had not begun, Peine would be one of those villages with 'only the old people left' and young people only there for holidays. El Litio, located a short distance from Peine, enabled the young to return and re-enliven the village. Ángel was particularly emphatic about the lithium mine having provided a flourishing intergenerational life:

> The school teacher, he was here for around 20 years, he had the idea that the Salar and El Litio was going to be something . . . By all means it is companies that have given a better life to the pueblo, which was *en el suelo* [at rock bottom]. By all accounts, the people who stayed here are fine. They are working, they have their things, their animals, their fields to plant. They have not lost *la onda* [the sense or the character] of who they were before, but they still have work and money . . . There is change, because, before when there was not this work, you lived with what you could produce here and that's it, and what you could obtain from the Argentinian side. When this work came, at least there was the possibility of working here and then merchandise came from Calama and what have you . . . and then this *pueblito* began to change.

Paulino, a man in his early 60s, also emphasized the positive economic change.

> When CORFO came, the pueblo, everyone wanted to work there . . . When CORFO arrived then things became easier and life began to change. In the '80s it was totally changed with El Litio. People began to operate buses and vehicles. I think it's good. If the mining had not come to this area, we wouldn't be here. Everyone would have gone to look for work elsewhere. There are shops here now, but before there was nothing. The company installed the plant here and from then everyone came to look for work. Everyone wanted to enter the company.

The beginning of El Litio represented the beginning of a local commercial economic life, the potential of wage labour without having to leave home, but still the continuities, the *onda* of before. It brought other workers to the region: men, who had money to buy consumer products and services, and women from neighbouring villages who, like locals, saw the promise of small business ventures in the village.

These memories contrast with local critiques of the lithium industry as it has grown. As Don Bartolo said, 'To begin with, the mining seemed good, but then opinions changed,' especially when there were influxes of *gente mala* (bad people) who arrived looking for work and disrupted life as it was lived. Earlier anthropological accounts of the region also track this equivocation. Blas Hidalgo's student thesis (1992), focussed on Atacameño culture in neighbouring village Socaire, includes the results of interviews with men who had moved to Peine from Socaire to gain employment.[16] He describes the perspective of a man from Socaire ('head of the household'), who worked in El Litio, and who 'did not miss the sacrifice of having to get up early to gather wood and return home very late from pasturing the animals or doing some other farm work' (ibid.: 285). Nonetheless, he said, 'neither during working hours nor during meals are there occasions for sharing time with co-workers' indicating wistfulness for agriculture's conviviality.

Some lithium workers I spoke to, like Don Angel, were proud of how they were able to be employed as young men at a low skill level and through hard work, learn on the job and rise through the ranks, without the need to gain formal certificates and training. However, not everyone was enabled to make progress like Angel. When Hidalgo interviewed workers in the late 1980s, one told him that he 'considered the company to pay low salaries to the workers and [take] advantage of the young people to abuse their services' (1992: 288). An engineer at the mine told Hidalgo that men from Peine and Socaire were employed if they passed a three-month trial, during which they would be trained on the job to undertake a range of tasks. He is quoted as saying,

> We put special emphasis on contracting *lugareños* (local people). Now, the training is something gradual according to the innate qualities of each person. Everyone starts with the same process as operators and from there one starts to see their skills and where they will go. There are *muchos tipos* (many guys) that are skilled enough to be mechanics, proficiently work with tools and capture the principal technologies, and there are others that plainly will have to drive a shovel, lift things . . . That is, you can't generalize about these people, neither Peineño nor Socaireño, each one you have to observe. They are good for some things and bad for others, so you have to look for the right place for each. It's all a cost, but if we see it from the point of view that you have to plant to harvest, it's also worthwhile, because people will become specialized and people will learn to operate the machines and this may be a model for those who come along, or perhaps their own sons, that way we have people of greater qualification and we don't have to bring someone from Antofagasta, or Calama, that will cost us 200, 300 percent more. And fine, the idea is to work with people from this place, it's cheaper. (Hidalgo 1992: 285–86, my translation)

The manager insists on the company's respect for local men, equalizing those he characterizes as deserving, who understand what it means to have become part of an industrial labour force, and those who already have training in its incipient technology. The company 'developed' (some of) the workers from the village, and the manager hoped this work would be a model for the next generation of local workers. This was good for business; a local workforce, *'lugareños'*, who cost them much less than bringing workers from elsewhere.

While some express nostalgia for a more gentle and collective life before, in 2011–13, those Amelia and I spoke to also emphasized how Peine, after El Litio, had gained some of the good things of modernity. The characterization of life 'before' as sacrifice and afterward, better, with work, was sufficient impulse to welcome the mining economy into the pueblo. Ever more men and women came to stay in Peine for the work, or the promise of work. A mixed economy was nothing new, it reaches back to the man from Peine who Philippi described in the previous century, working an artisanal copper mine with herds and fields to tend (Philippi 1860).[17] The history of labour migration and trade meant Peineños participated in the global economy starting well into the early part of the twentieth century (see also Vilches et al. 2014). Nonetheless, El Litio brought industrialization and wage labour to their doorstep with constancy and intensity. The salt industry was good for those men and women who gained from it. They could sleep in their family home at night and use their wages to improve the structure and furnishings of houses inherited from parents and grandparents, while also labouring with crops, herds, land, and water.

Don Geraldo, retired from working in the lithium mine, emphasized the twinned nature and duration of agricultural and waged work in the mine when he was younger.

> When we were working, we had animals also and we'd come home from work and feed the animals and then get up early in the morning and look to the animals before the workday. Everyone did this. Now the young people, no way! They get home tired [from mining or other wage labour] and that's it. So, things are a little different, they don't think about agriculture, *la pura plata, nada mas* [just about money]. Before, no, you had to work hard just to eat.

Old people everywhere make unfavourable comparisons of the present in relation to their memories of the past. However, the difficult question of how to manage a mixed labour economy was often present when I worked alongside members of a family as they were planting their fields or with people from different families with adjoining fields, as they stood and chatted while irrigating crops, or planted and weeded. Sometimes it may have been my presence and prior questions about change that were the impulse for such conversations.

One October, I was participating in the multiday events for one family's *siembra*. Gonzalo, who stands to inherit the fields and is responsible for their cultivation in the present led the work, with his three middle-aged siblings, his mother, one tottering grandchild and myself. Also working for Gonzalo were a couple who were trading a day's labour so that Gonzalo would work for them the following week and two men working as *peones* (paid day labourers). We undertook the tilling, weeding, and fertilizer-spreading for the handful of small plots and some of Gonzalo's sisters and brothers' wives prepared the large midday meal, supervised by Guiomar, the family matriarch.

The *siembra* was preceded by a brief *convido* and then another was performed by Guiomar before she instructed her daughter and I to help placing the seeds in the newly-dug rows. The exclamation '*Vito vito, cien porciento*' (referring to one hundred percent productivity) or '*en buena hora*' (lit. in good time) said loudly and heartily by all present, marked each *convido*'s end. During the work, short *descansos* were called by Gonzalo's older brother at (roughly) hourly intervals, during which beer and soda cans were handed around and some prepared wads of fresh coca to place in our cheeks. Each fresh drink or wad of coca leaf demands 'sharing' with the earth (spilling small amounts into the field, digging in a few leaves of coca) before serving oneself.

During one such *descanso*, Gonzalo's sister Amelia asked him why his teenage son was not helping the *siembra* in the fields that the young man would one day inherit. Gonzalo retorted, '*los cabros no sembrarán más*!' (The young guys will not plant anymore!). Gonzalo's son, he continued proudly, would be a professional, like his uncle. Agricultural life, he pronounced, was finished. Amelia responded, 'then why are *you* here?' and he retorted, 'my mother would die if I didn't plant'. His equivocal relation to his fields were played out in these scenes. Gonzalo showed off his newly acquired farm equipment and led a lively work party that sustained the '*alegria*' (joy) the earth demands for good results, and authoritatively proclaimed the inevitable end of tradition with his son's generation of professionals.

The texture of people's relations with the Salar and the emergence of extractivism through people's experiences with El Litio reveal important differences in local opinions about how to confront the near future. The mine's local bosses cultivated close relations in social and economic terms with Peine through employment, informal gifts, and occasional assistance in village petitions to a remote state. Company support for the village was an implicit recognition of the mine and workers' reliance on Atacameño relations with their water and lands, as well as people's willingness to work and their technical and ecological expertise, labour experience in other regions, and long-term relations with salt ecologies. The consideration of these material relationships would be a story missing many dimensions if it neglected to include an account of one long-serving manager of El Litio. This I do below, showing how interpersonal relations with company employees matter for understanding the history of extractivism.

Informal and Formal Intimacies

In 2010, Don Tomás, a poised man in his early 60s, granted me a short audience outside his home in the fading summer evening light. He leaned on the gate of his modern stone house, located across a small canyon from El Litio's mining camp, inside village land and with adjoining agricultural fields. Tomás and his Peineño wife tended corn and alfalfa, received their share of irrigation water, raised two children, and as landowners, and thus *socios* of the village and Junta de Vecinos, took part in everything from church activities to annual irrigation ditch cleaning and associated ceremonies. Respectful of community control over research, Tomás stated he would not allow me to interview him until I had received full *asamblea* authorization.[18] Nonetheless, he wanted to emphasize that, in general terms, the relations between El Litio and the village were congenial. As if on cue, Don Lando interrupted us, and the pair walked a few metres away. I waited while they had a conversation out of my earshot. Returning to speak with me, Tomás emphasized the 'mutual support' between Peine and El Litio. I asked him to tell me a little more and my notes of what he said reflect my sense that he was arguing for the kind of mutual reliance in the tenor of 'good neighbours':[19]

> When they began the mine, about 70 percent were workers from Antofagasta and about 30 percent from Peine. Now it is the other way around. The north American company had instructed him to think of himself like Santa Claus. He considered himself as giving the gift of work to the people, that is, teaching them how to work. During the previous year's recession when many other operations were laying off workers, Tomás took a cut from his end of year bonus and other luxuries, but the salaries of the workers from Peine were not cut. He described the many financial relationships between the company and village, such as the provision of electricity through the company's diesel generator (finalised in 2009). Through his wife, he calls the man who just came by, '*tío*' (Uncle), but everyone here, he asserted, is related and he makes his [non-local] workers have respect for the way of life in the village. When the company holds their Christmas party down at the camp, they invite the company workers and their children, but they also allow all the children from the village to participate. When the corn grows here it is beautiful in the fields, and people from the village have respect for what is in their garden and what belongs to their

> neighbours. Outsiders who work for El Litio are also taught respect, and he has control over them by having strict rules about workers having to stay in the camp and away from the fields. But the outsiders from other companies come in here and do not respect these things, take corn from gardens and drink alcohol in town.

Tomás' assertions reflected a sensitivity to the material relationship as one of shared economic wealth and sentiment; complaints about mine workers drinking and fighting and stealing corn from villagers' fields, and the north American owners of the mine envisaging his mythic status as giver of gifts.

In the days after our conversation, I observed Tomás' patron role in the festival of the Virgin de Guadalupe, celebrated each year throughout November in Latin America. In the local finale, the Virgin, having peregrinated around believers' homes all month watching human women eating *roscas* (donuts) and empanadas and drinking tea and hot chocolate, processes through the village accompanied by dancers and is received at the church. After a mass, a festive meal is served. In 2010, Tomás, his wife, the nuns and the current President of the Junta de Vecinos sat with the *anfitrión* (host of the event) at the head of the table. Later that same month, Tomás took a position with all the local authorities, for example, at a Christmas children's concert, when the Presidents of the Indigenous Community and the Junta de Vecinos, the Mayor of the Municipality of San Pedro de Atacama, and he, representing 'El Litio' stood on the stage.

After our conversation outside his house, Tomás invited me to take the employee bus to the Salar for a tour of El Litio in the next days. When I did, Marcos, from Antofagasta, who had worked at the plant for seventeen years, met me as I alighted with the workers from Peine. In a casual one-on-one tour of the mine (unthinkable since its expansion), he took me first to see the office building, the laboratory, the cafeteria, the bathrooms and (somewhat strangely to me at the time) the locker facilities. As we went from there to drive around the *piscinas,* sparkling white salt heaps, turquoise pools, and greenish yellow tailings of the lithium plant, he explained that his length of labour service to the company was unexceptional among their workforce. It was a comfortable job and secure. I was taken back to the offices to have an interview with Tomás, without him consenting to me recording audio. I scribbled my notes as he spoke. 'You see', he said, 'lithium does not contaminate the environment, not like copper. Our workers have conditions that are so much cleaner than other industries. We use only a little water, and the process is by its nature clean'. He showed me photographs from a slideshow (apparently created for the north American company owners) that celebrated El Litio's social investment in Peine and the other towns and villages of the Salar. He explained that workers receive interest-free loans from the company if they ask, he told me, since the workers may borrow in advance from the bonus they receive after each five years of service to the company.

The plant offices had the feeling of a family business. I wasn't 'inducted' into health and safety regulations, as was common in other mine sites (and as was the case when I visited the site in 2017 with a large group, after the Albemarle purchase) and Tomás and Marcos both told me they had no public reports they could share with me on social investment or other company activities. When I interviewed Marcos the following year, he also outlined many of these projects, but focussed on labour as the basis for good relations with the community. When I asked him if he thought this might change with the impending changes to national laws (and referred to Chile's recent ratification of C169), he replied,

> People work with us for many years and enjoy this relation that is *bastante amistosa, cordial y de ayuda mutua* [pretty friendly and cordial, of mutual help]. So the company feels happy with the people it has and the people are content with the place they're in, and this has been the case for 31 years . . . Things shouldn't change. Relations have always been good, friendly, and changes to the law are not going to change the people. We have always worked together, there shouldn't be changes . . . though one never knows what will happen.

The Peineño manager of the workers camp in Peine also emphasized labour and intimacy as the basis for good material relations. '*Todos se conocen*', he told me, everyone knows one another:

> Those from the company who work on the Salar have done so for many years, and for this reason we have good relations with the people of the pueblo. Why? Because they help the company and vice versa, they have helped the people of the pueblo a lot.

When I returned to Peine after my visit to the mine, I stopped in at Doña Alma's shop, and she exclaimed, 'What were you doing on the El Litio's bus this morning?'. When I told her that I had been a guest of the company and interviewed a very congenial Don Tomás, she retorted that he was *duro* (hard), with the workers. She listed her complaints about the way the company operated. Mateo, who worked for a few years in El Litio, expanded upon these comments after he, like Alma, wondered aloud what I had been doing on the mine worker's bus. Mateo emphasized a number of aspects of what he referred to as the 'irregularities' of the company.

> They say they adhere to safety, but this is thanks to the way that people from here in Peine work. We are good workers, safe workers, and we think about

> what we do because this is our home, not like those from Antofagasta who always make mistakes.[20]

El Litio was changing. Rockwood was taking over as operator and had plans for expansion in readiness for the forecast increase in global demand for Chile's lithium. Tomás would have to hand over the management of the mining plant to new men and new times. As he did so, the people of Peine remained both loyal and divided in relation to this man who had spent so much of his life embodying the company, as a boss, a relative, and a neighbour.

The ambivalence about impending changes surfaced in a personal exchange in 2012. I had volunteered with Paloma to serve drinks at a makeshift bar during a festival. Purchasing cans of beer for himself and other men, Tomás teased me, 'you must have fallen in love! Here you are again in Peine'. He and the men stood next to the bar, chatting and drinking. Gonzalo bought another round of drinks and raised a toast, 'To Don Tomás, who taught the *pueblo* how to work'. Gonzalo's cousin retorted that men in the *pueblo* knew how to work, but that 'Tomás' (pointedly without the honorific) 'expected Peineños to be *subordinados* (submissive)'. A heated conversation ensued, diffused by Tomás leaving to share a drink with others. Chatting with Gonzalo's mother some weeks after this event, she reflected on the unequal intimacies of the past, telling me,

> You know, I always defend El Litio. My husband worked for them, and my boys worked for them when they were first adults. When [her third son] was still a baby, my husband tried to start a union, and he lost his job as soon as the boss of El Litio heard about it. Soon afterward another boss in the company heard that he been sacked and got him a job again in the *piscinas* [evaporation ponds].

Good relations with Tomás meant village men could gain employment and the company enjoyed mostly compliant labourers, though there were intermittent attempts to form a workers union.[21] Tomás' imminent retirement was the face of change for Peine's relations with El Litio. Concerns about what this change would imply were debated in public meetings and around kitchen tables.

For over thirty years, Tomás was boss of the local workers in the Lithium mine and a *socio* of Peine's Junta de Vecinos. When I was present in the community in July 2011, he participated in urgent labour to deal with the frozen freshwater infrastructure as a result of heavy snow falls around Chakizoke (see Chapter 3). Tomás attended village meetings, assisted in manual work, and took a company vehicle with the small party of local and knowledgeable men to assess the situation at the site of the trouble in the mountains. Tomás was member of the local church and worker in annual irrigation ditch cleaning. Before the community signed the formal agreement with Rockwood, Tomás led the company's social investment in the town, performing the role of what his bosses characterized as a generous patron. He oversaw expenditure in education and health facilities, ensured the routine maintenance of roads, and contributed to provision of electricity. With the imminent growth of the company and its take-over by global operators, ambivalence about the advantages and disadvantages of the lithium industry on their doorstep became more prominent. Tomás represented the historical problem.

Ongoing negotiations between the company and the community about the 'shared benefits' agreement would replace the prior 'good neighbour' arrangement which relied on locals willing to work under company patronage. Sitting in Inés's kitchen in 2012, she began her observation to me much as usual, 'you know what, Sally? El Litio has been taking water from Tilopozo for I don't know how many years, and it is all free. They get everything from here all free'. The unfairness expressed in her comment reflects the sense that as 'El Litio', the company and the village had an intimate relationship, cemented by marriages and offspring among company men and local women, and some felt that the company had for too long taken advantage of these matters.

A Changing Salar

During President Michelle Bachelet's second term government (2014–18), recommendations from a National Lithium Commission underlined the need to consider lithium part of Chile's 'well-built future', one in which the mineral was strategic for green energy rather than military force, and would be part of 'sustainable development . . . [in which] an equilibrium between the social, the economic and the environmental' was sought.[22] A major concern of the Commission's report was over-extraction of saltwater by SQM.

The path to the place where Peineños used to harvest salt is marked by cairns of rocks and wonderfully preserved piles of mule droppings. When we went there in 2014, Don Geraldo and others indicated how the place was wetter than before and suggested that it was becoming inundated by freshwater. His explanation was that as more saltwater was extracted at the lithium mines in the near distance, freshwater surfaced here, wetting the salt crust and making it no good for harvesting. An increasing availability of commercial grain salt had replaced the laborious salt harvest, but this, he noted was not the point. Likewise, Valeria defined the reduction of people's access to the Salar's ecology, especially the salt lakes, in terms of an end to egg collection, saying,

> Now people don't go there anymore. And there are hardly any *parinas* on the Salar, not in Chaxa [Lake] or anywhere, or very few and not like before. CONAF banned the collection of parina eggs. But there are almost none to collect anyway, due to all the noise from the mining and *chupando la salmuera por abajo* [sucking brines from below] they are taking nutrients out of the water that the parinas used to rely on.

The collection of parina eggs was prohibited when the birds were made a protected species with the creation of the Chilean National Forest Corporation (CONAF) in 1972. CONAF's prohibition of entering the salt lakes to gather the eggs – the protection of vulnerable animal species – coincided with mineral exploration activities and eventual construction of the lithium mine. The flamingo eggs and feathers harvest, overtly connected to fertile relations with non-human beings such as spring waters and *cerros*, and salt harvest, focussed on sustenance and trading relations, were both ways in which people were active in the cycles of the Salar's ecology, its living, changing elements. One was prevented by law, the other ceased due to commodity replacement, both remain part of a distinctive Peineño knowledge of their territory and in the eyes of older Peineños, at risk of harm from expanding scales of lithium brine extraction.

In the life histories of older Peineños, lithium mining arrived at a time that is now marked by the beginning of *work* and the end of prior lives marked by *sacrifice*, but also tranquillity. Their stories also describe the effects of a long drought, a violent dictatorship bringing an end to work that straddled the border, and the imposition of state interests in the Salar as national territory and environment. These matters describe a temporal and geographical frontier of change into which the mine, El Litio, broke ground. While people say that the numbers of those cultivating fields in the oasis settlements of Peine and Tilomonte has declined and agricultural practices that were social as well as sustaining changed in intensity and character, these remain important to social, familial and local life. Work in El Litio is remembered to have generated flourishing in the village, with young people returning from the cities to live with families *and* get wage labour, and accompanying efforts at producing the comforts (and consternations) of modern life. Don Tomás' is a story of intimacy and local hierarchies that developed as he, manager of the lithium plant, boss of many men in the village, and husband of a local woman, became a patron of economic and infrastructural development. His retirement marked a number of important changes, including transnational capital takeover of mining company operations, and formal agreement-making, which also brought greater independent and state surveillance of workers and environmental conditions. Exploring these interpersonal, ecological and infrastructural ties between Peine and the lithium company describes some of the historical changes and continuities in mining capital and the expanding scales of extractivism as these enter the lithium boom of the second decade of the twenty-first century.

Notes

1. The Mother's Centre was part of Pinochet's conservative apparatus to gain networks of popular support among women in neighbourhoods and rural areas (see Valdivia Ortiz de Zárate 2010). Showing Inéz, this image, she reflected on a group of women daring to protest Pinochet's visit, doing a *cacerolazo* (banging pots and pans with wooden spoons).
2. The mechanics of this history are summarized in the Bachelet Government's Lithium Commision final report 'Litio: *Una Fuente de energia, una oportunidad para Chile*' (Comisión de Litio 2015) and reviewed in detail elsewhere (e.g. Barandiarán 2019b). For a discussion of the SQM mine, which established a workers camp in the village of Toconao, see M. Núñez (2002).
3. This may be translated as 'life was . . . more *fucked up, it was dire*, but the verb '*joder*' can be used in many ways, including as synonym for *aguantar*, to withstand (Dictionary Real Español. dle.rae.es).
4. For the sake of narrative clarity, these glossed-over corporate changes include the purchase of the Salar de Atacama plant by Germans Chemetall and its later acquisition by Rockwood Holdings, which began trading in 2012 as Rockwood Lithium.
5. These oral histories and recollections of the industry date back to when my interlocutors were aged 20–40 years.
6. I also draw upon an audio recording made in 2017 during a collaborative project of participatory mapping at the request of Community leaders. A detailed analysis of some of that material is published elsewhere (Babidge et al. 2019).
7. Don Bartolo referred to this as the *chururito*, or 'señaliza *del ganado*' (lit. signalling of the herds) in which coloured yarn is tied to herd animals' ears to attract fertility. Grete Mostny's (1954) work is detailed in regard to these matters.
8. The attraction of work on the railway was strong because it was only a couple of days' walk or bicycle ride from Peine and Tilomonte, enabling men and women to earn money and also meet agricultural, pastoral and other territorial responsibilities when necessary.
9. Various 'Yugoslav' families migrated to the region during the era of Nitrates mining in the nineteenth and early twentieth century and remained as traders and businessmen. The wonderful collection of testimonial-style personal histories, '*Así Pasa Cuando Sucede*' by González, Selti, López and Puca and collated by Cecilia Uribe (2014), contains similar accounts of Atacameño people working for Yugoslav families in the region.
10. A swift and early act of repression was regime's murder of union leaders and members in the north's copper mines, especially those associated with the state-run Chuquicamata.
11. In 2016, cleaning out the church storeroom before it was repaired, we found one such old rifle.
12. The construction contractor, 'Termonor' began operating at the same time as 'Foote Minerals', the manager of the company said proudly (when he gave a short interview in 2013). He had arrived at the same time, and has a well-established dormitory-camp in Peine.
13. Chile's national holiday. September is also early in the planting season.

14. 'Pucha' (also *Chuta, Chucha*), expression or exclamation signalling exasperation.
15. '*Vai pa*' in Chilean Spanish. *Vaya por*, or 'go' (imperative of the verb to go), '*adentro*' 'Inside' the Salar, or 'go to work'.
16. Some of Hidalgo's fieldnotes held by his supervisor, Prof. Maria Grebe, and have been digitized. They are available on the University of Chile digital catalogue catalogued under Grebe's name.
17. In Catherine Allen's work on coca (2002), she points out that subsistence pastoralism is a kind of misnomer because cash was always needed to meet people's desires for a good life, such as festival materials, or for metal tools, and for many other things.
18. At that time, the Community leaders (the Directiva) had signed a letter permitting me to talk with whomever would agree to do so, with the proviso that I could not use any data collected until a full community *asamblea* approval was forthcoming (see Chapter 1).
19. I jotted notes as he spoke, with some verbatim phrases, then typed up an account once I returned to my rented room, which appears here edited for clarity.
20. In a published conference paper (Babidge 2011), I reported Mateo's and others' criticisms, as well as the positive opinions of El Litio that people told me at this time, using a pseudonym for Peine and all personal names. After providing a copy of the paper to the Antofagasta-based SCL manager (as agreed with Don Tomás when we spoke), he rang me to angrily inform me that he would no longer make it possible for me to speak with company staff.
21. The formation of a workers union was something local men initiated as soon as the company changed ownership.
22. See final report (ibid., page 6). Quote of Bachelet is from media source of the launch of the lithium governance policy, 26 January 2016 (see www.gob.cl/en/old-news/president-bachelet-chilean-lithium-will-not-be-a-case-of-frustrated-development-but-an-example-of-a-well-built-future, last accessed 20 April 2020).

Chapter 6

MAKING RELATIONS

Intentional Intimacies of Advanced Extractivism

Escondida and Peine were early to adopt formal contractual arrangements over community 'development benefits'. However, some mining company employees from other companies operating in the Salar neither envied, nor had their company emulated Escondida's approach (at least not for some years afterward); least of all Rafael, then manager of 'Environment and Society' of the lithium miner, SQM, who was responsible for the Annual 'Sustainable SQM' report. Over the course of three field visits I had pursued him by phone and email for an interview and had badgered the SQM field officer to help. Rafael eventually granted me permission to audio record a short conversation in a hotel lobby on the edge of the city of Calama. On the recording, Rafael's antagonistic tone and my coughing and sniffling due to a cold I'd developed on the bus journey from Peine to the city, make for a sense that *neither* of us wanted to be there. I asked him to describe SQM's relation with communities of the Salar, and he complained that changes to national environmental legislation (in 2010) had generated 'uncertainty' for their company, particularly since Peine's community leaders were using the expanded regulations for Indigenous consultation regarding environmental impact prior to SQM's planned plant expansion to push SQM for a contractual agreement. SQM had a policy against signing community-level agreements, he informed me, because these, he asserted 'trade' environmental approval for community benefits. He criticized state process, other company's practices, and community desires, and joked that in the 2007 agreement signed between Escondida and Peine a global giant had been 'duped by a village'. He said,

---

Notes for this chapter begin on page 175.

> Escondida [*laughs through his nose*] needed to achieve the environmental procedure [for the project 'Pampa Colorada'] – *that they did not achieve* [tapping the table] – they attempted to transact environmental effects through an agreement, and they were left with an agreement and no environmental permission! So, the discourse that annoys us, *alot*, is that which speaks of a relation. Because the social license to operate must be provided by the communities and the environmental license must be provided by the authorities. So, what happens? You [the community] give the social and environmental license away when you ask for an agreement and then shut up [*callarse*]? But the project may be bad, environmentally speaking. Or vice versa: the project might be really good and socially dubious [*muy cuestionado*] . . . . And look I'm not saying that we're the angels of the story, right? But in this locality [Peine] and others we have gone and told them what we do, we have reported to them, and they understand us perfectly well. And if they have questions, we have tried to clarify these as much as possible. . . . [And] when it comes to it, we can help in many ways. We cannot do anything if they put these barriers of having to make an agreement with a final number, you see? That is a direct support and nearly blind, ehm, and we are losing, in my opinion, opportunities that could be developed that have been successful in other places, real help.

To Rafael, agreements are instruments in which communities receive development benefits in exchange for being silent on concerns about environmental harm of industry.

In a business sense, the trade looked to Rafael highly ineffective. He scoffed at corporate competitors who had not achieved their extractive objectives despite having gone to the trouble of making an agreement with Peine. Rafael asserted that SQM operated with transparency, with an openness in company reporting, responding to questions when asked by community members and providing 'help' to communities, all without a formal agreement. However, not long after our conversation, SQM was found guilty by Chilean courts to have over-extracted brine and freshwater in the Salar de Atacama, and guilty of not reporting their knowledge of environmental harm, and also guilty of corruption (cf. Comisión de Litio 2015). According to the regional association, the Council of Atacameño Peoples (CPA), SQM has inadequately informed communities of environmental harm over many years.[1] The result of the legal proceedings was that SQM was ordered to pay out millions of dollars in compensation to the communities of the Salar de Atacama, and to do so via community agreements with each of the Salar's Indigenous communities, brokered through the state body, CORFO.[2]

The Peine community's history of making agreements with Escondida (see Chapter 4) shows that neither agreement "shut them up" about their concerns regarding ecological and social impacts from groundwater extraction. The community has not acted as though signing an agreement is purely transactional, nor acted in such a way as to signal that the agreement resolves their relationship with the mining company. In Canada and Australia, and increasingly in Chile and elsewhere in the world (O'Faircheallaigh 2023), Indigenous peoples have engaged the technical instrument of agreements as mechanisms to fund Indigenous livelihood projects, gain local employment, seek mitigation or compensation programs for harms to territory and impacts on society by mining, and for participation in environmental monitoring and management to militate against impact on Indigenous territories. Unlike the promises the SQM manager outlined, such as remaining transparent or to listening to community concerns, agreements are legally enforceable.[3] Where an agreement is formed through the negotiating parties, with the company and community members having input into its design, many Indigenous communities have judged these to be adequate or that the agreement may generate potential for good. Since the late 1990s, the Peine community has sought to draw mining companies into agreements with them, as one way to gain a legal form of recognition (which is not entirely forthcoming from the state) about the fact of company activity in their territory. The payment of financial benefits to the community under the terms of agreements is significant to local wellbeing and wealth, but as I have shown in the case of Peine and Escondida, agreements do not directly or necessarily transact economic benefits for silence on environmental harms. An agreement is almost never, counter to Rafael's assertion, a 'final number', or simply the terms of a 'transaction' for either party. Agreements create new relational conditions and these too are far from unproblematic. As critical analysts and practitioners in industry have pointed out (Owen and Kemp 2012), agreements often mark a stage in an evolving relationship. This may continually bristle with advanced extractivist uncertainty or conflict, rather than achieve or establish concrete terms between the parties, as is implicit in Rafael's assertions.

The agreement, as a legal instrument, is part of the globalizing moral economy of advanced extractivism, and arises from the imaginary of rationally acting economic actors who as parties to the agreement abide by its terms. As an instrument unmediated by the state, it also provides for contingencies that are otherwise unavailable in the local material (economic and ecological) relations between companies and communities. The agreement instrument holds a promise to make explicit and tangible a relation that is implicit by the fact of territorial intervention of mining and largely symbolic in company policies of CSR. Furthermore, while community members may be employees of the company and thereby have knowledge of certain operating realities or have experience of being subjects of company gifts or local philanthropic spending, meetings and other events centred on an agreement potentiate communication with company bosses at close range and on matters specific to the good of and conceivable harm to territory and people. The terms of an agreement give rise to interactions between community leaders and company managers, relations in space and time where the details of company activity can be addressed.

For mining companies, agreements demand relationship management. Escondida (in 1997 and 2007) and Rockwood Lithium (2012) both signed community-company agreements with Peine, and when doing so employed professionals whose job it was to be the personal and personable connection between the parties. Anthropologist Marina Welker (2014) writes about mining company employees such as these 'enacting' the company and Alex Golub examines the way in which such employees 'personate' the company (2014) in their relations with communities and civil society. As these and other scholars of mining worlds have shown (Kirsch 2014) interpersonal relations, those between social relations managers and community members and leaders, can tell us much about the broader inter-entity phenomena of extractive 'encounters' (Owen and Kemp 2017; Bainton and Owen 2019). Community relations employees of extractive capital have the job of ensuring that company-community agreements are enacted, made, or avoided according to company policy, operationalizing company Corporate Social Responsibility (CSR). Their role is in communication to community members about environmental impact reporting and negotiating with community leaders about the implementation of social benefits projects, but also in managing everyday activities that companies undertake in the village or community surroundings. They are often, but as I demonstrate in this chapter, not always, the first point of call for community members when seeking to bring the company to account or seek outcome on an issue.

The company social relations employees are responsible for securing in interpersonal relations what a paper agreement may not have achieved but makes possible on the horizon, a mundane sense of good and socially responsible business as usual. I refer to their work as 'making relations' and use the phrase to consider the moral and material dynamics of advanced extractivism at this intimate level of intra and inter-personal. The fact that large-scale mining pairs ecological exploitation and CSR (with key terms 'sustainability' and 'responsibility') is an ongoing contradiction about which I argue we must be more than simply cynical. The point of my focus on employees of companies is not to criticize these individuals, but to consider the ways in which persons from company and community experience their activities, especially in, around and related to agreements, as a field of relations, and thus a moral field. I provide insights into how the fundamental contradictions of capitalist extraction and 'good business' also troubles company actors responsible for being 'good', and what this (further) reveals about the fundamental inequities of advanced extractivism.

Rafael from SQM (quoted above) was reluctant to be interviewed, rude and dismissive. His character provides a stark counterpoint to Escondida and Rockwood employees who were mostly generous with their time and in my experience, interested in making *good* relations with Peine community members. I had the occasion to observe interpersonal engagements between Escondida's employees and the Peine community and between them and Rockwood Lithium (Albemarle) employees, and to some extent SQM's while staying in Peine during different periods of fieldwork, and I also travelled to cities in the region to interview company employees in regional offices, and at conferences and meetings of mining professionals. Other chapters in this book draw on the incremental ethnography of many visits over a decade of fieldwork, but this chapter is necessarily constrained to earlier periods of the research endeavour.[4] The community relations employees I refer to in this chapter represent company conduct in relation to the Peine community during a specific time period. I begin by examining how, for Escondida employees, their failed project for water extraction, 'Pampa Colorada', kept haunting interactions even after 2007, and revealed 'trust' was unresolved by making relations. The second part of the chapter focusses on Rockwood's first community relations employee (after 2013), who relies on the history of intimate ties between that mine and the village, the mutual reliance and patronage (as outlined in Chapter 5), to emphasise the idea of 'shared responsibility' for the environment. I then return to the question of persons as agents of broader company strategy, using Escondida employee musings on their role to critically consider the interpersonal dimensions of advanced extractivism.

Relationship as Intention and Violence

Nicolás had just been appointed manager of Escondida's small community relations team,[5] when he agreed I could record a conversation with him in the company offices, in the northern neighbourhood of the regional capital city, Antofagasta (see Figure 6.1). After pleasantries, he opened up a 'PowerPoint' slideshow on his desktop, swivelling the screen toward me and clicked through a corporate-branded presentation. It depicted 'before' and 'after' images of housing and sports stadiums the company had renovated or rebuilt and photographic evidence of completed social projects in the city of Antofagasta and in Puerto Coloso, the fishing settlement adjacent to Escondida's private port infrastructure. He saw potential to do good work also in Peine. He said,

> My vision is to be able to forge a *convivencia armónica* [harmonic coexistence] . . . to find respect in a mutual appraisal and to collaborate in sustainable development as a progressive process, that will also increase the social license so the company can operate . . . [*He asked rhetorically*] What does it mean to have harmony? [*He answered himself*] That we have mechanisms for resolving conflict, not that the conflict doesn't exist, conflict can exist, but we can control it and we can talk about it in a regulated way and within determined parameters. [We want] to achieve this with the communities, this harmonic coexistence, principally founded in respect between us and a mutual evaluation [*valoración mutua*]. Respect implies many things; it means we have to provide clear information. Achieving harmonic coexistence means we must have a bigger presence [in the Community], stable and uninterrupted. For this reason, Luchita[6] is working there, we have to have a closer dialogue, more direct and more stable. With this we can forge a coexistence that is more harmonic.

As Nicolás outlined, the express focus of work in the Peine community was to ensure Escondida's extractive activity operated not just smoothly, but in harmony with society. This language seems hyperbolic against the practical examples he gave; to inform the community in clear terms about company activities and position a company fieldworker in the village to better manage company communication with community members. In some ways, it is this mismatch between what is said by company employees such as Nicolás and what the company achieves that frustrates Peine community members. An example; the members of a group of older women weavers who agreed to being photographed after Escondida gifted them a new loom, found that the image appeared in glossy company brochures about social investment, reappeared in the popular press, and on the company website. The disjunct of small acts of company generosity against the 'global giant' making a fortune from extracting water from Peine's territory sat beneath constant pressure by community leaders on company employees to address matters of importance.

That community members judge company gifts and recompense as incommensurate with extractive harms is one aspect of what Nicolás, and other representatives of the company, sought to change through engaging interpersonal relations to soften the edges of a remote, contradictory, and powerful entity. As discerning judges of character, Peine community members appreciated the well-meaning company employees with whom they interacted, and made explicit differentiation (in conversation with me) between liking such persons and opposing the actions of the company. However, all Escondida engagements with communities in the Salar were haunted by their failed water extraction project 'Pampa Colorada' and it

Figure 6.1. 'Minería Enfermería'. Geoglyph-style graffiti on the hillside facing Minera Escondida regional offices in the northern suburbs of Antofagasta, 2009. © Sally Babidge.

was Nicolás' job to avoid a future in which the company did not get their projects approved by the government authorities, which meant appeasing the community. He told me,

> After Pampa Colorada, the company learned, had to have learned, that it has to have a greater closeness with the communities. Fundamentally, it [the company] thought that because Escondida is very far away and doesn't have people nearby, next door, on the borders of the mine . . . [he trails off]. But in the end, you are sharing the same territory, a territory that for many years they have had important use of. And not to repeat a process that they might see as a process of recolonization . . . such as what happened many years ago when the Spanish came and they, the different *pueblos*, submitted under arms. The idea is that they don't look at us like we came here and used their riches without their permission. Now Escondida doesn't do things like that. Escondida is in line with the law, to all the formal norms that this country has, and it has gone through a process of foreign investment and it abides by all the aspects of the law. But one thing is what the law permits and the other is that one must work with the Community who is close-by, and this comes before the Chilean State.

Nicolás's explanation displays a sense of critique of colonial history; he uses this logic to assert that the relation the company has with community is not neocolonial, but a primary, 'sharing [of] territory', which 'comes before' the relationships with the state. Escondida's activities, the mine itself and their groundwater extraction to the south of the Salar pre-date Chilean state regulation prescribing consultation with Indigenous peoples about impact. The fact that the Escondida mine lies some 100 km from Peine, Nicolás admits, was added impetus for the company to avoid dealing with Communities of the Salar on the impacts of their operations. He omits the fact of the company's misleading scientific reporting about groundwater extraction impacts in the southern Salar. Instead he focuses on the more visible Pampa Colorada project, in which the company proposed to extract water from the high Andes. While groundwater at high altitude can be just as enigmatic as other vertical topography, the aquifer subject to Escondida's proposal in Pampa Colorada was 'upstream' from the Indigenous communities whose lives are in and around the foothills of the southern Salar, and adjacent to national conservation reserves, which heightened the public sense of potential harm.

In representing the good work of the company, Nicolás makes implicit reference to voluntarism. On another occasion, he had asserted that the company does 'more than the law requires' in engaging with communities, and he shifted his reference to legitimized company exploitation of land, not taken under duress. When we spoke in 2011, Escondida's policy in regard to communities was in a process of change, responding to altered Chilean state legislation and associated measures of environmental regulation and Indigenous consultation. Structurally, Escondida's community relations were part of the area known as 'Health, Safety, Environment and Community' (HSEC) part of BHP Billiton's global operations. While 'community' was always directly associated with the transnational's corporate policy, for Nicolás it signalled a global perspective on *Indigenous* community recognition parallel to Chile's ratification of C169; Escondida's approach is a step above actual Chilean law and local practice. Nicolás gestured at how BHP as a global giant was more aligned than other operators in the Salar with international instruments and global corporate responsibility discourse in the recognition of Indigenous rights in territory, water and self-determination.

Nonetheless, projects identified in Peine's Development Plan that were supposed to be funded under the agreement with Escondida had stalled. This presented a problem for Nicolás and his team, who were responsible for ensuring the company's social investment could be actualized. As Nicolás says, he employed a 'specialist in community relations', Lucha, who was expected to be involved in making sure the projects funded by the company would be carried out, and her position was designed to produce outcomes for the company's HSEC division. Lucha had training in working with 'the poorest communities' in Latin America, Nicolás told me, so would be the ideal 'person on the ground ... prepared to get as close as possible to people, with a simple and understandable language'.[7] His strategy for communicating was to reverse the idea that the company was an unreachable monolith and he hoped that Lucha would be able, he said,

> [N] ... to co-manage and have more involvement and participation in the [Development] Plan, and when we have more participation, more involvement, more *engagement* [said in English], to generate more confidence and become closer. This is what she means to achieve ... this is what Lucha will seek to achieve there [in Peine]. A year ago, there wasn't ... we were very far from that, so ... I have confidence that we can achieve these processes in that we can improve the relation with the Community.
>
> SB Sure. In the past the relationship was not very...?
>
> N: Yes, due to the Pampa Colorada project ... yes. That project collapsed principally due to a community issue [*un tema comunitario*]

If the company had undertaken greater steps to make relations during Pampa Colorada, he argued, they would not have had the problems that resulted in region-wide Indigenous and environmentalist resistance. He seemed to assume my co-understanding with him regarding cause and effect of this event, as well as invite me to see him, as a representative of the company, to be transparent in talking about the company's activities, and in this case, its mistakes.[8] However, the failure of Escondida's Pampa Colorada project exceeds the failure to gain Indigenous peoples' approval. Broad opposition by communities throughout the region brought the circumstances of the Pampa Colorada project to regional government and national attention; Indigenous communities, non-Indigenous residents and tourism operators of the Salar, and regional and national environmentalist groups formed a loose coalition against the proposal and other extractive projects in areas of ecological (and thus tourism) interest (see Bolados 2014). The project was ultimately rejected by the government body on grounds related to inadequate measures in Escondida's planning for mitigation of environmental harm.[9] If trust is key to making community-company relations, then the Pampa Colorada provided a significant breach.

At the end of 2012, Nicolás again agreed to my request for an interview at Escondida head offices in Antofagasta. He began by announcing that Lucha had managed to do a lot of good work in making relations with people in Peine after one year in the job. The process was slow, but he hoped she might be able to live in Peine and set up her own office, ensure the notices of meetings and other kinds of information were properly advertised, help the community with all kinds of small projects, and train community members so the company might 'improve capacity'.[10] Nicolás' idea about 'capacity building' was directed specifically at community administrative and decision-making processes, which he judged deficient. The community needed to be more organized and efficient if they were to achieve development.

During a fieldwork visit the following year it seemed that Nicolás's plans were unravelling. One evening, the president invited me to sit in on a meeting with Escondida employees Nicolás and Lucha. With their lawyer, the *directiva* had made a formal written complaint to the head office of Escondida's operating company, BHP (in Australia), incensed about employees' strategies to get close to community members. Nicolas and Lucha had not confined themselves to assistance with development project management, but instead had interfered in community politics and decision making. The result of this was that Lucha was prevented from moving into Peine on a permanent basis, and she was accused of becoming too close to one community leader, and of seeking to influence that person's decision making, and thereby creating divisions among community members. This interference in community affairs caused anger and a sense of betrayal among those who had worked closely with Lucha and Nicolás, temporarily shut down relations between company and community, and emphasized divisions among community members. One community member became estranged, and conflict simmered between families who had taken sides. Nicolás' project of making relations with the community through Lucha, as local instrument, put to work the rhetoric of good will, friendship, and mutual interest in 'community development benefit' for the structural ends of mining business outcomes.[11] Nicolás' strategy not only failed in business terms, but it also had violent outcomes, generating bad feeling among people and catalysing community division.

A Fragile Mutuality

In contrast to Escondida's distance from the village and workers, the lithium mining projects on the Salar are both visible and familiar. As outlined in a previous chapter, Peineños have long laboured in the evaporation pools and laboratories of El Litio (now Albermarle), and a good number worked as contractors to SQM Salar. Both operations are well known. In 2012, community leaders invited me to observe a meeting between them and representatives of Rockwood Lithium, in which they discussed the impending agreement being negotiated between the community and the company that promised a formalization of long-term 'mutual' interests. Before the agreement, Rockwood relied on a small handful of employees to undertake the work of communication and relations with the village. These were men who had worked in the company for a long time, risen to management of the plant, and some have familial relationships with community members. Two such men, now retired, were pivotal; Tomás (see Chapter 5) and Enrique, an engineer who had been employed in the company since its inception.

During a break in the proceedings of the meeting, Enrique and I stepped outside for a brief interview. He told me that since 1984, the company had been investing twenty million pesos (CLP, roughly USD 30,000) each year in Peine. He listed on his fingers as he spoke to me – they have lent machinery, done road works, improved the school buildings, provided a dental chair. 'We have always supported the village', Enrique told me, but this assistance had been 'all very disorganized'. In 2007, he said, the company had begun talking about a formal agreement, and in a way that reminded me of the conversation with Nicolás, Enrique asking rhetorically, 'Why did we do this? We did this for three reasons . . . '. Rockwood sought an agreement with Peine as a way of managing the company's internal process, second, to respond to impending legal changes coming with C169, and finally because they wanted to expand their processing plant. For this they would need an Environmental Impact Assessment report and undertake extensive community consultation. At that time, it was not necessary to do Indigenous community consultation, Enrique assured me, since their plans to expand the mine slightly predated new legislation, but they would not prejudice (*'perjudicar a'*) the community just because the company had lodged their EIA a few months before the law changed.[12]

Enrique and I were standing on the patio outside the building designed as Peine's Guest House dining room, which was being used at the time as a kind of community boardroom. The building sits high in the village, with an impressive view out across the Salar to the dirty white lines of lithium evaporation ponds on the western horizon. I squinted into the distance and asked Enrique whether he thought the company's relationship with the village was changing with the new formal process of agreement negotiations. 'Look [*Mire*]', he said, he had worked on the Salar since 1976 and he knows Peine. He gestured to Mateo, then community president, listening to our conversation, and addressed him, 'you must have been running around then in short pants'. He continued 'I care for [*tengo cariño a*] the pueblo'. He has, he said, respectful relationships, and the company had never discriminated between workers from Peine or workers from other parts of the region in terms of labour conditions and benefits. In my memory, at this point Mateo walked away from the conversation.

Contemporaries of Mateo and older men complained to me that they are offered only low skilled work and that the company treated local workers differently to those from outside, denying them opportunities provided to other workers. 'Fair dealings' was not an adequate representation of local opinions on the worker-boss, or community-company relationship. Indeed, Mateo had opened the meeting that day by stating that they needed to settle the terms of a formal agreement in order to 'manage the future and release the hierarchies of the past'. Thinking of his words, I asked Enrique whether the new formal agreement would see the employment of a community relations team. He replied that between Tomás and himself are, 'executives with a relation to Peine, and [we] transmit the responsibility of family to the company'.

Both Enrique and Tomás, who represented company patronage of the village since the 1970s, and in Tomás' case was literally family, were retired within the next year. These individuals whose presence and life histories represented significant embodied hierarchies – in which, for example, adult men and elected leaders of their community may be treated in openly patronising ways by company managers – were no longer at the helm, and Enrique's appeal to a sense of shared personal history necessarily shifted. Nonetheless, a mutuality of the conditions in the remote and marginal area of Chile was a rhetorical current in the new generation of company relations employees.

After the formal agreement was finally signed, Rockwood hired Rubén, a journalist by profession, to work on the Company's communication with the community. I met Rubén in November 2015, when I begged a lift with him from San Pedro de Atacama to Peine.[13] Over the course of the hour's drive, Rubén agreed that I could audio record our conversation and during the following year's fieldwork, he also agreed to be interviewed, inviting me to sit in the office of the mining camp to do so. He lived in the mining camp during his shift, along with the other Rockwood employees, and complied with the strict codes the community had instituted in regard to company employee socialising with community members outside meetings and formal events.

What was it like, I asked him, to be part of this shift from the old company relations to the new? Enunciating each word slowly in turn, Rubén said 'it's – a lot – of work. It's a lot of responsibility, not just for us, but also for the communities'.[14] The Peine-Rockwood agreement is based on 'shared value'; it is an equity deal in which the company agrees to distribute to the community a percentage of profits rather than a fixed annual amount. Gundermann and Göbel (2018) argue that the deal marked a break from reputational activities of other companies in Chile. A movement toward such agreements is ever more common as a trend of 'inclusive capitalism' (Dolan and Rajak 2016, see also Langton and Mazel 2008). Mateo told me that he thought the idea of an agreement with Rockwood based on profit share was generated by the community and their lawyers. He argued that given recognition of Indigenous territorial rights and self-determination, they asserted that the royalties paid by industry to the state (to CORFO) should instead go directly to Indigenous people on the basis of territorial sovereignty of the Salar. The notion of 'sharedness' in the agreement may be understood to partly arise from the territorial and resource mutuality in which the mining company and the Peine community rely on and potentially profit from the Salar.[15] The appeal to mutuality is also rhetorically positioned in commonalities of worker experiences on the Salar in the extraction and processing of saltwater in shared industrial and ecological history. However, the terms of sharedness in these relations and experiences were easily unsettled.

Since the years of CORFO in the late 1970s, and the beginning of 'El Litio' in the 1980s, extractive operations and workers have relied on much the same territorial, water and material infrastructures as the community itself. Rubén explained to me this 'special' relation between Peine and Rockwood as manifest in the shared experience of daily living:

> If they lose electricity, I too lose electricity. If their water is cut off, then for both of us it is cut off, if there is wind[16] then this affects me too. I am in the same place. This has meant that the Rockwood people, those in this camp [*tapping the table where we sat*] have a special sensitivity in regard to Peine . . . To me . . . the big difference in regard to Rockwood and the other companies is that Rockwood has a camp in the pueblo. The problem [for the company] is that suddenly from the Community's perspective the relationship may not seem very beneficial. We drive our vehicles [on their roads], we're using water, we're using energy . . .

The relation between Rockwood and Peine as mutually dependent on emplaced waters, infrastructure and the climate tied them to the sense of one another as entities through time with interlinked economies. The bodies of workers and managers in the camp have relied on Peine's domestic infrastructure, and the company and its employees were involved since the late 1970s in the provision of that infrastructure and its ongoing maintenance. The mutuality was material and substantial, mediated by bodies, territory and water.[17] No other extractive company in the Salar has such a relation. However, this sense of mutuality was precarious.

The dormitory camp, built on land that had, many years before, been excised from Peine (as a locality) by *Bienes Nacionales*, is situated looking over a canyon and lower sector of irrigated fields. Some thought it had erased a part of the Inca Road and most saw it as within Indigenous community territory. Rubén agreed,

> It is in their territory . . . *bueno* [*he leaned forward and spoke directly into the recording device with a grin*] 'according to Chilean law the territory here [*the mining camp*] belongs to Rockwood' [*laughs*]. Yes, look, we use their services, we use the streets of Peine.

Aware that national Indigenous territorial rights legislation was changing, Peine's leaders had begun to challenge the inconsistencies in prior legal grant of land title to the company for their camp. Rubén's playful irony indicated his own knowledge of the structural inequalities that underlie the partial nature of Indigenous territorial 'recognition', in this case; exempting land that was excised for the camp well before Indigenous Community territorial registration. But his discomfort connected to other forms of resource mutuality.

By law, the company must provide 'potable water' to its workers. However, when Rubén and I spoke, Peine's leaders still refused to add chlorine to the water from Chakizoke (see Chapter 3). Legally, the water the company provided to their workers was not 'potable'. When I asked Rubén what Rockwood was doing about the water situation, he whispered, 'at the moment we are in infringement of the law'. The precarious terms of the informal relationship had been possible for many years; this tiny operation, extracting a 'strategic' but small-time mineral, situated remote from cities and regulators, with its managers including those willing to share the marginal resources of business and life with Peineños. Rubén feared that Rockwood would have to close their camp, not because of minor illegalities, but because the times were changing.

> I might be mistaken, maybe nothing will happen, and the camp will continue to be here. But I think also that one has to realize that a big part of the strength of the relation that Rockwood has with Peine is exactly because Rockwood is *here*.

Rubén understood that the sense of sharedness was fragile. With every household in Peine having at least one person employed in El Litio at some point since 1984, and now younger people gaining work in Rockwood, there were relations of many kinds to manage. Rubén had faced formal accusations of intervening in community business and a period when they refused to speak with him or the company. Rubén, sitting at the cusp of change, articulated the strangeness of old relationships with their bonds of patronage, and the ways in which such bonds had begun to dissolve with a generation of mining men retiring, younger community members who felt less beholden to the company becoming elected leaders and challenging the company to long-held property, and the new owners of the company demanding 'global standards'.

While Enrique somewhat de-emphasized the fact, the agreement that the community and the company signed in 2012 was an important factor in Rockwood gaining government approval for expanding the mine using only a Declaration of Environmental Impact report, and not a full EIA.[18] The company-community agreement had been made in terms that reflected an old way of operating, but now Rubén and others must try to make it work in a new regulatory regime, a new atmosphere. He gave me an example of how things had changed. In 2015, four members of the community leadership, led by a younger generation, had been willing to spend a whole day in Antofagasta with Rockwood employees discussing the agreement and its function in order to address how to meet their contractual obligations, especially regarding environmental matters. Rubén said, that in 2012, when the agreement was signed,

> the new environmental processes that exist now did not then, there was no oversight ... Now, in the agreement there is a promise to form an environmental monitoring council, which has still not been done. ... The agreement with Peine secured the company's EIA [*sic.*], so the authorities will come along to check up and say, show me where this [monitoring work] is! And if I don't have it, they will hold me responsible. So, we did this day-long workshop in Antofagasta and the directiva attended.
>
> SB: Oh, I think they told me about this, but not with much detail.
>
> [R] Really? It seems this had more importance for us, I guess. I found it spectacular because we found some agreement on environmental responsibility, we made a proposal ... ehm, that is, I think that this also shows something of the maturity of the relation [we have] with Peine.

It did not immediately occur to me why Rubén found the whole process 'spectacular'. For some years, community leaders had talked about their desire for a program of environmental monitoring in the southern Salar, around Tilopozo and Tilocalar as well as the salt lakes, and so I had not considered the meeting so unusual; however, it did demonstrate a level of willingness by both company and community to work on the project. This, he envisioned, would become 'a new and first level standard' for any agreements that would be made later by the community or other companies, and Rubén seemed disappointed in the apparent lack of shared excitement.

That Peine and Rockwood are in league with one another, Rubén considered, extends from the village out to the Salar. Leaders have articulated a range of concerns over the years regarding the impact of mining activity, especially saltwater extraction, on the salt lakes and their surrounds where *parinas* breed, though until more recently, most of these concerns have been directed at SQM and not Rockwood. The salt lakes on the southern end of the Salar are protected ecosystems and the community has tried for many years to have rights to and responsibilities for these areas recognized.

For my first five or so years of fieldwork visits, the lakes had only been part of historical narrative or a sunset sensation, the expanses of saltwater in the near distance picking up the colours of the sky. I had sometimes asked those I was with at this time of day, 'could you take me there?', or 'when do people visit?' The mining company restricted access, I was told, though others said visiting was possible. Some people had not been for many years, since mining began and CONAF banned flamingo egg collection. By the time I saw the lakes up close, when I tagged along on a trip with visiting scientists in 2016, there was a chain and lock across the track that led there and the community *directiva* held the only key. Gleefully, Rubén said he and the *directiva* together decided to lock up the salt lakes. This meant that Rockwood's commercial competitors, SQM, indeed any company or individual (such as researchers or tourism operators), had to seek permission to enter and collect the key from community leaders. Given SQM's environmental monitoring program included entering the lakes on a regular basis, I witnessed a number of tense phone calls and in-person visits between community leaders and SQM employees about the barrier and its key. Eventually, a process for access was settled. Rubén recounted the 'keys to the lakes' event to me as a kind of stunt played on SQM, Rockwood's corporate competitors, one in which the community and he were co-collaborators. However, the whole process carried stress and an administrative burden for community leaders, for whom managing access to the lakes is a responsibility to them taken deeply seriously.

Rubén's sense of the 'shared' aspects of the company and community future has a basis in the terms of mutual material existence and life histories, factors that partly influenced the creation of the formal relation in the agreement. Nonetheless, the assertion of mutuality entails an elision of old hierarchies, and of the inequalities of impact for shared decisions, even those as apparently minor as locking a barrier. Sharing, as a technical term and ethical assertion, is inconsistent with the responsibilities involved in looking after life in place into the future. Rubén was sympathetic to this knowledge, reflexive about the conditions he worked in, and in our conversations expressed discomfort about ongoing inequalities for local employees of the mine before he departed from the company the following year.

Company Agency

Field operative employees, Lucha for Escondida and Rubén for Rockwood, were tasked with implementing company funded projects, being agents in company 'good business', as well as brokers expected to both gain trust of community members and deliver outcomes for the company. In both cases, employees' activities had been implicated in community tensions and outright conflict, to which the community responded by sending letters of complaint to company head offices. As Rubén reflected,

> Perhaps for the community it was more important, or for the people of Peine it is much more important, to have formality in the process, and we, perhaps were coming from a perspective of informality and closeness that we had with the community before.

Indeed, one of the emerging principals in this period by which Peine's leaders began to assert control over relations with outsiders was through insisting on formal terms of operation, and largely rejecting the mode of operating that I have termed 'making relations'.

Nicolás and a member of his team, Santino, told me in 2012 that there was a new valorization of community relations in mining companies in Chile, so that the job was gaining some 'honour'. Previously, Nicolás had confided that, as a devout Catholic, he saw his work as 'doing penance'; while he believed his work was for good purposes, his career would probably suffer.[19] His characterization of the communities and causes that received 'social benefits' from Escondida was marked by the ways in which he articulated the moral responsibility of people like himself. In candid discussion, he reflected on being pained by community behaviour he found abhorrent (in general, what he saw as disorder) and on numerous occasions asked me to help 'them'; once he meant the company in terms of improving community relations, but on other occasions he suggested I assist the community leaders. He also couched much of the company's work that he oversaw as voluntary work to better the lives of 'the poor' and to help people progress, revealing an older ethic of business patronage and philanthropy, quite in contrast to the business case of his company's public documents about CSR.[20]

With Nicolás and Santino, we discussed the impending changes to the Environmental Law (2010) in Chile, meaning the process of community consultation would become necessary for all new developments. Santino said that he thought the consultation process was an 'opportunity' to develop better relations between companies and communities, but they were not sure how it would be applied across the Salar. I asked them whether their senior managers in the company considered the impending consultation regime to be 'an opportunity'. They told me,

> Nicolás: The company knows that this is a critical issue . . . the company is scared of consultation. Or rather, not scared of it but that they are getting involved in . . .
>
> Santino: uncertainty.
>
> Nicolás . . . the company functions a lot with themes of risk, and business is evaluated in terms of risk. They like to work with maximum possible certainty or control of risks. They know that this is a zone where it is difficult to have control, not even 70%.

Santino's role, he said, was to try to deal with this uncertainty for Escondida. He sought to get them to a position where the community might work together with them on environmental issues. He said 'I want to see the company, more than simply *entregar* [pass on] the environmental reporting information *compartirla* [to share it]. That all of the community can understand it', and he explained a desire to implement an education process in order to 'build community capacity'. This would be a tool for the community, he continued, not only in dealing with Escondida, but also with the other companies. One aspect of pre-regulation company engagement with community, concerns the ways in which employees' personal ideas about creating a sense of shared purpose in order to undertake such 'capacity building' caused problems rather than solving them, as signalled by community leader complaints about company interference in local business. Nonetheless, there was a sense that the consultation landscape was about to change.

The energy among industry employees for considering different ways in which they may engage with communities are a stark contrast to the fatigue of communities who must respond to consultation. The burgeoning interest in the Salar with the lithium boom coincided with these significant political implications for Atacameño communities represented by the CPA, especially Peine. The following chapter examines such matters.

Conclusion

Mining company employees' agency, their relations, actions and discourse, disarticulate 'the mining company' as monolith, as well as create and recreate 'the company' in space and time (cf. Welker 2014; Golub 2014). Tracked through time and associated with particular events, employees' commentary and personal views indexed where they did not directly refer to the dynamics of national events, the shifting global standards on corporate 'best practice' developed in corporate think tanks, and the changing priorities delivered to them in corporate budget allocations. In problematic ways, individual employees' work hitches 'corporate good practice' to what is temporarily possible in the constraints of corporate policy. In conversations with mining company employees, they revealed to me how relationships between their company and the Peine community were sensitive to time, scale, and ever-changing company structure and policy. An examination of their efforts to make relations shows how employees' conduct often overlooked the sometimes radically divergent interests of their employers and the community, as well as the fracture between company driven ideals for community acquiescence and the community relations employees' ideals.

Ethnographic research often contributes to critiques of political economic processes by showing how seemingly abstract systems are enacted by persons of particular social groups, as subjects of and agents within such conditions. Community relations employees are employed to both see the terms of a company-community agreement are adhered to and go beyond them in the interests of the company, operationalizing aspects of the company's Corporate Social Responsibility (CSR) – an umbrella term for the projects and policies on which extraction companies may rely for their claims to be doing good business. They do so by interpersonal actions of 'doing [company] good'. Therefore, do companies claim to be 'doing good' through enacting social investment (or 'development') projects, or does planning 'environmental monitoring' and other forms of participation stand up to the claim of doing good?

Making relations is an advanced extractivist activity by company employees who are charged with enacting company claims of doing good business. Two main effects of this are discussed in this chapter. The first is that the interpersonal attempts by community relations employees to develop personal relationships with community leaders and key community members may be considered unethical or ineffective. My critique of this social practice is that 'making relations' presumes that human mutuality, a sense of interpersonal respect in the course of conversation and 'getting to know one another' may be scaled up; that, for example, (community) person to (company) person friendliness may deliver community-company trust and deliver outcomes favourable to the company. On the basis of the examples narrated in this chapter relating to Escondida, 'making relations' significantly underestimates Peineños' expertise in managing advanced extractive relations and fulfilled its potential for betrayal and harm to individuals and social disruption to the internal relations among community members.

The second aspect of making relations considers the metaphor of mutuality, 'sharedness' activated in this process through appeals to experiences and life in place as well as discursive appeals to development desires. In the case of Rockwood, co-existence of the mine, its camp and the community over time and in place allows for convincing appeals to the sense of a past and future together, though these rely on the informal patronage and personalities of old. The dynamics of what has been mutual in these relations – a history of work and long-term influence – also underlies something of how Rockwood and the community have come to see one another as having shared responsibility for environmental harms and hazards of company activity.

Extractivism, the systematic support of the growth of extractive industries at the cost of people and environments, promises economic development benefits to communities and elides the potential for ecological harms. In advanced extractivist conditions like those in Chile, companies and communities negotiate this elision in relations framed by company-community agreements. In doing so, they declare that the world can tolerate both rapacious development *and* environmental health through mitigation measures; that is, through 'good relations' with Indigenous communities and 'sustainable' resource extraction.

In Peine and elsewhere, people use the relationships of contractual agreement to grasp hold of capitalist actors who they are otherwise unable to control. Employees of extractive capital seek to manage communities through 'making relations'; interpersonal and everyday human exchanges at close range. The problem is that this form of 'relations' may be inconsistent with company activity in broader territory and over time.

Whatever feelings arise between persons (a mutual sense of good intentions, for example) does not translate to more systematic relations as *good*. A key example is Nicolás himself, widely considered by people I spoke to in Peine as a 'good man', while Escondida's activities as a whole have been consistently judged as intolerable in community decision making. That is, relations with company employees are fragile, since all are aware of the teleology at their core and the transcendent calculations of having 'done good' found in company brochures continue in the face of small conflicts. Such matters support rather than challenge critical understandings of the potential for advanced extractivism's notion of 'good business'. In an examination of the everyday ethics of encounter, the interpersonal and small local actions of good faith, there is a temptation to see the possibilities that making relations is ethical business. While communities can gain access to financial, infrastructural and other kinds of benefits through closely-negotiated deals 'in good faith' with mining companies, in the balance of things, the result of making relations always serve the interests of capital. The question of greater good – to which companies appeal in their arguments for continued business – is difficult to sustain. Impacts matter at scale and through time. Families are riven by conflicts seeded by company activities, and these tensions threaten community endurance in the conditions of advanced extractivism.

Notes

1. Interviews and conversations in 2011–12 with Community leaders in Peine and also Socaire are peppered with accusations regarding SQM. These developed into legal challenges. These and an Environmental Tribunal case are discussed in Chapter 7.
2. At the time of writing, this process remains contested by the CPA and individual communities, since the agreement was wholly designed by lawyers acting for CORFO, who (the CPA and communities charge) engaged insufficiently with communities themselves.
3. As a legal instrument, the agreement may be also used by companies or the state for leverage in regulatory processes. As a lawyer who worked on an agreement made between Rockwood Lithium and Peine in 2014 informed me, when they made this agreement, 'SQM accused Rockwood of buying the community and increasing the standards of industry, but I defended what we did because I have a conviction that it is right'.
4. Given published critique and my involvement in community affairs after about 2016, it was no longer tenable to seek interviews with company employees.
5. I refer to this team as 'community relations' as a gloss. Escondida/BHP periodically changed the structure, range of functions, and personnel who were employed to manage relations with the state and other companies, Indigenous and local community liaison, environmental management, and media communications. I refer to 'community relations' as a shorthand for the changing group of employees that worked on mostly community, sociocultural and environmental projects and 'external affairs'. During Nicolás's term of management, his team included worker health and safety.
6. Like all personal names in this book, this is a pseudonym. I have used the diminutive version of a common name to mirror the original speech.
7. There is a gendered dimension to this relational work that others have noted (Gajardo 2021).
8. One may argue this is a kind of 'impression management' (see Carrasco 2010, 2020), or perhaps what Welker (2014) calls 'attaining faith', in which company employees assert truths that align with company policy. To me this leaves aside the question of whether people 'consciously' acknowledge the contradictions of the position they assert. Either way, Nicolás, as I show, was considered a nice person by people in Peine, but still an employee for a company that continued to seek to extract more water from their territory.
9. The project sought license to extract groundwater for ten years from a protected high altitude wetland ecosystem and one connected to Indigenous community water sources, those on which they depend for agricultural and domestic purposes. Detailed accounts of the project's failings can be found in the government authority's report (Comisión Regional de Medio Ambiente 2007).
10. 'Capacity building', as one of the many managerial languages of neoliberal development embeds assumptions about community deficits in existing skills, approaches, and knowledge (cf. West 2016).

11. There is a gendered component to these events. A full analysis in those terms would reveal detail of a complex series of structural and interpersonal matters. It is worth indicating that women employees in and as Indigenous community 'targets' of CSR in Latin America has been the subject of important feminist analysis (e.g. Billo 2020; Gajardo 2021). The personal implications of community leadership for Peineño women I spoke to indicate that local gendered dynamics should also be considered. However, I cannot address these matters without breaching the privacy of individuals involved.
12. I note that Peine's leaders had told me that *they* had been seeking a formal agreement with El Litio for some years. The company were aware of the delays that the community could cause if it were not consulted: it was not voluntarism on the company's part.
13. Peine's president that year had suggested I do this, since the regular bus had stopped running daily from Calama and Rubén was often coming and going.
14. In our conversation, he referred not only to Peine's relationship with the company, but also the agreement the company signed in 2016 with the regional Council of Atacameño Peoples, representing the interests of eighteen communities. I refer here only to where he discusses the company relation with Peine.
15. In a brief interview in 2015, the lawyer involved in negotiating and drawing up the agreement also emphasized this ideal of 'sharedness' that dominated the negotiations and the formation of the ongoing *mesa de trabajo* (working group) responsible for the institutionalization of ongoing relations.
16. Wind (and dust) storms on the Salar and in the Atacama Desert are often fierce, resulting in low visibility and (often) damage to power infrastructure.
17. Men and women from Peine describe the wages earned from the mine as having improved their homes (literally money for tiling floors, installing kitchens, adding rooms or installing tin roofs). Family is also a felicitous and substantial dimension of the material facts of mutuality that exceed 'community relations' discourse. Many welcome children have been born of relations between women of the community and men employed by or contracted to the lithium mines.
18. These reports are publicly available online. see DIA, '*Secador Planta Potasio Rockwood Litio Limitada*' 2013 Available at seia.sea.gob.cl.
19. In objective terms, it did not. He moved into a high ranked appointment in public relations in the regional offices of a utility corporation, and later with another large mining company.
20. Rajak's (2011a) work, especially chapter 6, details this rhetorical transition and the persistence of company patronage as the moral tenor of relations.

Chapter 7

THE OVERBURDEN OF PARTICIPATION

At an opening plenary of the third international industry conference on 'Social Responsibility in Mining' in Antofagasta in 2015, a senior official from the business group the Chilean Minerals Council emphasized company transparency, saying they must now operate according to the 'mantra; that which I must hide, I cannot do' (*'lo que tengo que seconder, no puedo hacer'*). The conference brought together mining company employees, consultants and researchers within the industry, and representatives of government agencies to discuss the policy and practice of CSR in Latin America and beyond.[1] Many were focussed on the environmental regulations that President Michele Bachelet managed to pass in the last months of her first term as the nation's President, and which created the Ministry of the Environment and the Environmental Assessment Service (SEA, under Act 20.417, 2010, modifying Act 19.300, 1994). Following Chile's 2009 ratification of the International Labor Organization's Convention 169 (C169) on the rights of Indigenous peoples, the new environmental mechanisms required that companies undertake Indigenous community consultation on all new developments where they may 'impact' the rights and territories of Indigenous people. As Miguel had foreseen (Chapter 4), this brought Atacameño Peoples, mining companies and the state, with increasing frequency and intensity, into practical and political engagements with one another.

In the north, the new legislation was just beginning to take effect, and the novelty was causing a buzz among conference delegates. A copper company employee reported with enthusiasm on a community relations program, which 'promised great benefits'. They were 'only doing *sondaje*' (test drilling), an activity 'below the threshold' of the law for undertaking consultation, but it was *'mal negocio'* (bad business) if you don't consult. At the final plenary, a company representative on the platform spoke about

---

Notes for this chapter begin on page 201.

the 'radical change' of the new regime of Indigenous consultation, mirroring the analysis and enthusiasm others conveyed. Sharing the platform with representatives of the state and industry was president of the Council of Atacameño Peoples (CPA), at the time leading negotiation with the lithium extractor, Rockwood/Albermarle over a 'shared benefits' agreement, which had attracted a good deal of public attention (O'Faircheallaigh and Babidge 2023). The senior bureaucrat from the state agency for mining commented that 'social considerations' should not hinder economic growth. For his part, the copper company chairman opened by saying, 'I will speak from my heart' and opined the value of 'shared objectives' and 'development together'. An audience member asked speakers about changes in the 'just distribution of resources' and another expressed their hope for a 'rethinking [of] the relationship with the Other'. Such intellectual interest, even excitement, among industry representatives and consultants about potential socio-economic change associated with the new government regulations contrasted with the subdued countenance of the CPA President. When he eventually spoke, he emphasized that the agreement the CPA was negotiating with Albemarle had some potential for advancing community self-determination ('*autoejecutación*'). For them, lithium was not 'mining riches', but part of the territory and ecology of the Salar. As such, the communities must manage mining companies' power.

The comments reflected public comments by each of the Salar's Atacameño community leaders, who at the time expressed disquiet about multiple new mining proposals arising at the same time as the regulations came into effect necessitating Indigenous community consultation. The Peine community and their territory was arguably at the centre of this upsurge of consultation and negotiation. For the Peine community, who for many years had demanded a voice in 'development' processes and recognition of territorial rights, the new regulations meant that they were finally procedurally included. They were also beset with company proposals: both copper miners, Escondida and Zaldívar, sought to extend their groundwater concessions in the Monturaqui-Negrillar-Tilopozo aquifer and lithium extractors SQM and Albemarle expanded their annual volumes of brine and freshwater pumped from the Salar, a small copper miner had been hustling to break ground just a few kilometres from Peine, and new lithium concessions threatened on the horizon. In addition to these major projects, a continual stream of state agencies, mining company representatives, exploration outfits, tourist operators, researchers and media jostled for Peine's leaders and community members' attention, asking to schedule meetings or run workshops. Faced with these multiple and mounting pressures to which they were bound to respond, leaders of the Community of Peine made an administrative complaint to state environment agencies in which they encapsulated the complex problem of extractive projects in their territory as a '*sobrecarga*'; quite literally, an overburden.

Alongside Peine's leaders' analysis of extractive overburden, critical commentators turned their attention to the socio-environmental impacts of expanding lithium extraction accumulating against pre-existing environmental effects from the copper industry (for example, Gajardo and Redon 2019; Barandiarán 2019b; Liu and Agusdinata 2020; Bustos-Gallardo, Bridge, and Prieto 2021; Jerez, Garcés and Torres 2021; Voskoboynik and Andreucci 2021s). The Salar de Atacama and the potential environmental impacts of the lithium boom has attracted interest in the Salar from across the world.[2] Popular tech journal *Wired* featured a story about stunning aerial photography capturing the undeniable visual impact of extractive processes in the Salar (Katwala 2019), and the BBC ran a story featuring Atacameño people titled 'The farmers who worry about our phone batteries' (Livingstone 2019). Subsequently, research articles written by civil society, citizen science and the popular scientific press mushroomed.[3] The administrators of the website of US Government Geological Survey chose to highlight satellite imagery showing extensive changes in the Salar's surface owing to the growing number of lithium extraction plants between 1993 and 2015.[4] A Reuters journalist who has filed a number of stories on lithium and the Salar since around 2016, reported in February 2020 that German carmakers were calling for greater Chilean governmental environment oversight of the industry, given that 'residents and environmental groups worry about potential damage to a regional ecosystem home to an ancient Indigenous culture, lagoons inhabited with rare flamingos and a booming tourism industry' (Sherwood 2020). Far from simply 'damaged by' the lithium boom, as this journalist and others implied, Atacameño community leaders, were, as they have done for many years, actively dealing with lithium companies and other extractive industries in their midst. Nonetheless, conditions had intensified.

With the lithium boom, the Salar became a mining frontier of a more recognizable kind, drawing critique from environmental groups, researchers, regulators, and other capitalist publics of twenty-first century extractive industries. The lithium boom is accompanied by public interest in ecological justice. Alongside Chile's 2010 environmental regulations was the creation of Environmental Tribunals (beginning in Santiago in 2012). The Primer Tribunal Ambiental, named for its jurisdiction over the northern regions of Chile, began undertaking proactive labour and prosecuting cases of environmental damage in 2017. The Court has become important to the politics of the southern Salar. To add to these national dynamics, international interest in transition away from fossil fuel energy and toward renewables has pointed to the need for socially and ecologically 'just' transitions, while industry remains dependent on groundwater and minerals such as lithium extracted from Indigenous territories.

As an ecological crisis has surfaced with the lithium boom, it has brought new sets of threats, as well as possibilities, an apparent historical aperture in which we may ask – can these conditions be transformed? (Roitman 2013) Chile's President Boric, elected in the wake of the *Estadillo Social,* addressed the *political* crisis with a commitment to systemic change. Boric promised an environmental Left: social reform that would also protect the environment. This promise can be seen in the Lithium Strategy (Gobierno de Chile, 2022a), which sets out a massive public investment in and partial nationalization of lithium resources, manufacture and research, alongside the creation of protected areas and a technical research institute situated in the north. While electoral outcomes of the northern regions show a strong majority supported change represented by Boric, the Lithium Policy has not been unequivocally celebrated among Atacameño leaders in the Salar. They have questioned the logic of environmental measures alongside the mandate for massive extractive expansion, and both taken part in and challenged negotiations over partnerships with lithium miners.

Atacameño people are not simply *subject to* conditions described above, the moral, technical and political economic complexities of advanced extractivism. Their responses 'trouble' it (Haraway's 2016). An inherent and unavoidable relation with the Salar – relations to a material politics that include an earthly multiplicity – may be detected in the dynamics of Atacameño participation and response. As such, Indigenous practice neither fits the imaginary of Indigenous people as easily aligned with environmentalist stewardship or those building resistance to all mining, nor does it match industry expediency. Instead, the partiality of their participation unsettles and shifts advanced extractivist expectations.

This chapter discusses a series of events in which the Peine Atacameño community, as polity and incorporated body, both participated in and refused the (in)convenient ideological positionings of non-Indigenous politics, as well as the administrative expectations of a non-Indigenous technical and economic rationality. They insisted on maintaining relations with their own kin (of the human and non-human kind) as key to defining good practice, moral authority and the future. For these reasons and others I outline, the effects of these actions constitute a slow resistance to advanced extractivism. In what follows, rather than join the (welcome) chorus publicizing ecological crisis due to the escalating extractive industry, I outline the less-recognized aspects of what Peine's leaders characterized as 'overburden'.

Participation as Overburden

'Overburden' has parallel meanings in English and Spanish. It is the verb meaning to carry too much, in reference to a physical or symbolic load. Overburden is excess, disproportionate and surplus to capacities. Chile's environmental regulations, taking effect at around the same time as an intensification in extractive pressures, entailed considerable administrative labour for communities since the burden of examining, analysing and potentially contesting company claims falls on their shoulders. Participation in the process is also morally weighty, with leaders and community members having to judge potential harm to territory, ecological commitments, and potential socioeconomic benefits, and do so under the gaze of a global and national public. An account of these material and moral burdens together narrates what I heard as an Atacameño analysis of the excessive conditions of advanced extractivism in their protest about 'overburden'.

Overburden as a noun refers to the overlying rock or soil removed in the process of extracting an orebody; an obstacle to the accumulation of extractive profits, set aside in mining business-as-usual. Critics of the effects of community participation in extractive governance, the inclusion of communities in the process of in extractive contexts, may well consider the metaphor: communities are 'extractive subjects' (Fredrickson and Himley 2020), drawn into participating so that their concerns are registered and then set aside in the steps toward licit accumulation. Thus, in their survey of EIA outcomes, Barandiarán and Rubiano-Galvis (2019) showed that communities in Colombia and Chile used C169 to demand participation in environmental governance, and that once such matters were heard, courts largely upheld extractive 'rights'. Participatory mechanisms may proceed entirely legally while promulgating environmental injustice by actively excluding diverse sets of people and ideas as simply not part of the designated process (Bustos, Folchi and Frangkou 2017). While inclusion of communities in extractive governance has re-politicized some dimensions (Merino 2018), the incorporation of Indigenous Peoples movements into political process may be 'less disruptive than social protest' (Falleti and Riofrancos 2019: 87). Participation may be both 'a route to liberation' and a tool for 'co-optation, sometimes a practical problem [or] ethical injunction' (Kelty 2017: S77, cf. McQuarrie 2013). Thus, the outcome of legal recognition of Indigenous rights to be consulted over environmental impact has been analysed as a deepening of community involvement in extractive governance, with multiple outcomes, often as agitator-within (Acuña 2015; Falleti and Riofrancos 2019). Converting Indigenous demands to participation in extractive governance is about order, mundane administrative and technical activity that positions the state as responsive to social movement demands to address historical and actual injustice (see Goodale 2016). Drawing communities to participate in governance in this way may also limit their capacity to enact a full range of responses. Whatever the political outcome, the systematic inclusion of Indigenous peoples in extractive governance, through state regulations on consultation, entails considerable effort in Indigenous responses.

In 2017, Joaquin and I sat across a desk from one another in the community president's office discussing the preliminary stages of community consultation regarding Escondida's Monturaqui EIA. He expressed fatigue in the face of the many loose ends and other unresolved mining proposals, the local tensions among members of the Peine community about interpreting social and environmental impact, as well as the sometimes-divergent interpretations between Peine and other Atacameño communities of the Salar. Behind Joaquin was a whiteboard wall calendar for the month of November showing 21 of 30 days marked with at least one meeting. I expressed surprise at this visual excess and he laughed and said, 'That's not even all of them!' (and agreed that I could take a photograph of the calendar, at Figure 7.1). Items on Joaquin's calendar included two different monthly company-community agreement governance meetings (called '*mesas de trabajo*' or working party meetings) with the lithium miner Albemarle, one for the Peine-Rockwood agreement and the other for Joaquin's involvement in the CPA agreement with Albemarle.[5] On the calendar was marked an upcoming meeting with a gold mining company; two community meetings about internal matters; meetings in at least six different regional locations associated with the ADI or with inter community business; appearance and helping to run a 48 hour event to raise money for people with disabilities; meetings in the village for local education, health and tourism committees; and meetings with at least four different government agencies. Two days before our conversation, colleagues and I ran a participatory mapping workshop that Joaquin chaired, along with community leaders and a dozen other community members, thereby adding to his agenda of meetings. Like many of the other meetings, our workshop was not brief: it had a starting time of 8pm to accommodate people's working lives, and when we finished after midnight, Joaquin and others spent the next few hours preparing snack boxes and painting placards. The following morning, fifty or so community members (and I) took these to Antofagasta on chartered buses for the purpose of spending an afternoon outside the government agency SEA. Our noisy protest accompanied a meeting of members of the regional government voting on and voting down a proposed new lithium mine on the Salar.

Figure 7.1. Joaquin's calendar of meetings. © Sally Babidge.

Why are these meetings such a burden? It is not just the length of time they take up. As outlined in Chapter 1, the phrase '*la asamblea manda*' (the *asamblea* decides) refers to the moral imperative for the *directiva* to defer decisions to a meeting of members of the Peine Atacameño Community. There through discussion and consensus, or by vote, a position is reached on matters of importance, especially those in regard to territory and water, and relations with other Atacameño communities and organizations, with mining companies, the state, and those who seek to work with or write about the community (such as researchers). *La asamblea manda* thus expresses the authority of the community as a whole in contradistinction to the *directiva* (the elected board of community directors), who are considered not at liberty to make such decisions themselves, especially in regard to responding to requests by outsiders of the community, *on behalf of* the community. Given the imperative to take decisions in *asamblea*, the president and members of the *directiva* may thus suspend all external demands by deferring to the need to take these to an *asamblea*, which may not occur in time to provide a response. Alternately, in order to make decisions in response to multiple external parties, the *directiva* may feel pressure to call *asamblea* more frequently than the members of the community would like. In its capacity to delay events, the moral ideology of *la asamblea manda* has the potential to provide an ethical and temporal leverage for community in relation to (more) powerful outsiders.

Like any polity, the *asamblea* is dynamic, marked by Peine's system of roughly annual elections and is thus as unpredictable as any polity, its flux representing changes in active members over time and the political mood according to the social and material conditions of the present. In the many *asamblea* I witnessed, those who expressed their opinion and spoke up, informing others and influencing decisions, were often older (aged 55+) men and women. In recent years, seniors have begun to send other members of their households to the *asamblea*, changing its majority composition to those middle aged and younger members. With mine workers on twelve-hour shifts, and many working age men and women employed in the lithium mines, associated contractors and service industries, people are short on time and collective decision making takes time and willingness. When not working in the mines, people's crops, animals, domestic labour and family responsibilities occupy the day. To accommodate community members' labour demands, regular *asamblea* (between 2009–19) were scheduled for 8 p.m. With a late start, and discussion and consensus decision making having primacy, they then stretched long into the evening, sometimes well past midnight. People fell asleep or, to some amusement, began snoring in their chairs, exhausted from life otherwise.

Like those before him, Joaquin as community president was doing his service to the community, helping to uphold his responsibilities regarding Indigenous territorial authority, without salary or personal recompense for the role. He, as others before and after his term, spent significant time and effort organizing *asamblea* responses to extractive and other proposals. Many of these involved long technical reports about which few have a good grasp of detail and few are pre-circulated, but over which, when summarized and presented in *asamblea*, there are considerable differences of opinion. Such matters were increasingly on the agenda. However, getting a quorum at an *asamblea* could be a problem. As a forum in which industrial activities and proposals are considered and decided upon, member disapproval of previous or potential decisions may be expressed in non-attendance. To help address the technical questions that needed answering and the mounting workload, the *directiva* used funding gained through mining negotiations to employ a person they referred to as a 'Technical Secretary', and then, after some years, they employed Alejandro, an Atacameño man with a degree in environmental engineering and his origins in a neighbouring Salar community. Alejandro's job included fielding company and government requests for consultation with the Community, monitoring company activity, finding professionals to undertake analysis of DIA and EIA reports lodged in relation to Peine's territory, liaising with lawyers and collating responses to extractive activity as directed by the *asamblea*, as well as translating technical matters into the terms to be best understood by members of the *asamblea*.

With these technical demands, Joaquin is among peers as younger members begun to take on the burden of political service to the community. Don Benicio, then in his late 70s, recounting to me his own period as an elected village leader around the mid-twentieth century, gave an example of how things have changed. Today, he said, community leaders' time and efforts are taken up primarily with outsiders, especially mining companies.

> El Litio [industry] came and they just installed themselves, they didn't ask for permission from anyone. They only recently did the paperwork for this . . . For the water . . . The same with Escondida. They had to look for an agreement . . . then afterward the leaders practically only dedicate themselves to work with the companies, make agreements or have meetings. Before, no, before the leaders didn't do this. These days they must dedicate themselves to supervise the company as well.

'Supervising the company' conveys Benicio's evaluation that community leaders must be attentive first to the activities of mining capital, who previously began their mines without having to settle anything in writing with the Indigenous landowners, and have only recently 'done the paperwork'. Some seniors like Benicio remain active in *asamblea* if their physical capacities permit, proffering their long view into community discussions of the sometimes-inscrutable present. Many have stood aside as a new generation of leaders are elected by the *asamblea* to deal with the technical and political complexities.

Given the decision making process that the Peine community follows, in which they must carefully consider industrial proposals against community and ecological futures and respond to company claims, participation and consultation is burdensome in technical terms and also socially exhausting. With the new environmental regulations coming into effect and the demand for lithium rising, both SQM and Rockwood (Albemarle) lodged their proposals with the government authority SEA to expand their extractive activities and meet the coming boom (Mejía-Muñoz and Babidge 2023). Supported by state regulators at that time, both companies maintained that their expansion activities would not have an impact in regulatory terms on Indigenous territories.[6] As outlined above, the lithium proposals arose amongst multiple others for ever greater extractive access to the Salar's earths and waters. Below, I detail the Peine community's slow resistance to these advanced extractivist dynamics.

Slow Resistance to the Scale of Lithium Mining

The mine SQM Salar has increased significantly in size since its beginnings in 1995 and has operated in ways that have allowed it to underreport the environmental impacts of its operations. This situation has frustrated Atacameño peoples in the southern Salar, as they have watched SQM's plant fill the horizon with dust during the day and light at night. In 2019, SQM Salar reported that it extracted up to 1500 liters per second (l/s) of '*salmuera*', groundwater carrying salts, from the nucleus of the Salar (SQM 2019). The brine is pumped from beneath the surface and into ponds that expose the water to evaporation. When the brine in the evaporation ponds reaches an optimum concentration of lithium, it is removed and transported to industrial plants close to the city of Antofagasta, where lithium carbonate and lithium hydroxide are produced. The whole process takes between twelve to fourteen months. SQM states that 'brine is not water' since it has a high concentration of total dissolved solids (around ten times higher than that of seawater).

Lithium production demands the extraction and evaporation of these forms of groundwater for obtaining the final products (potassium, lithium carbonate, lithium hydroxide). A small amount of water and mineral is recovered after processing that is reinjected into the salt flat. The company also uses freshwater to keep the industrial process lubricated. According to public record, SQM has the right to extract 240 l/s of fresh groundwater (and Albemarle has the right to 24 l/s). The freshwater quantities are lower than those extracted by copper miners. However, SQM and Albemarle's mining concessions allow extraction of large quantities of brine. In her studies of the public documents, scholar Barbara Jerez (2018) estimates this at around 185,100 cubic meters per day and freshwater quantities that are estimated at 22,800 cubic meters per day. Given that brine is not regarded as 'water' in the Water Code (1981), lithium companies do not require water rights for brine extraction, making this dimension of their extraction invisible in terms of whole of basin water management and environmental regulation in the Salar de Atacama (Babidge et al. 2019).

Unsurprisingly, SQM's assertions about 'sustainable' extraction have been increasingly contested by Atacameño peoples, especially the CPA; however, there has been no halt to SQM's extraction. In 2018, the company trumpeted its approved expansion plans, which entailed an increase to 2.2 million tons annually by 2030. As ecologists noted, the 'brine reserves [of the area] are undetermined', thus it remains unknown what is 'the reasonable or sustainable level of brine sucked under the new expansion agreement' (Gajardo and Redón 2019: 4–8).

In 2019, the president of the CPA, then from Peine community, along with other CPA members travelled to Madrid where the UN Climate Change Conference (COP25) meetings had been re-located from Santiago.[7] During the inauguration of the Chile pavilion, they confronted Chilean senators and political representatives on the nation's climate change policy. The CPA President pointed to lithium, powering 'green energy' for Europe's richest economies, but which relies on exploitation of Indigenous territories. In part, the President of the CPA said,

> Today, [the government] have come a long way to present a discourse here in Madrid without having taken any real action in the Salar de Atacama. The President of the Republic himself, who has called on lithium companies to keep exploiting the Salar knows that today there is very little knowledge of what state it is in. Today, the State has not done what is necessary, and they are drying out the salt flats . . . [8] (President of the CPA, 2020).

Discussing these politics one afternoon in late December 2019, over a beer or two in Mateo's patio in Peine, Francisco began sketching the 'Robotic Parina' (Figure 7.2), a flamingo with a switch marked 'SQM' and 'CORFO'.[9] The rise and rise of Chile's lithium industry on the Salar de Atacama would happen at any cost, he commented, enabled by the state. Indicators of salt lake surface degeneration and reduced reproductive success among Andean flamingos have many people worried, and while measures of ecological protection of fragile saline ecosystems are part of the new Lithium Strategy (Gobierno de Chile 2022a), extraction continues alongside.

Aspects of SQM's activity have been questioned since 2012, when SQM was embroiled in a case of corruption associated with the state's granting of a new production license to then-leading global producer of lithium. The company was under the spotlight for a number of years, subject to investigation under re-elected President Michelle Bachelet's Commission on Lithium (Comisión de Litio 2015), which recommended that the Chilean state emphasize the 'strategic importance' of the industry to the state by strengthening regulations around its extraction, production of energy and ownership of the industry.[10] In 2016, the Environment Superintendant (SMA) oversaw a legal charge that the company SQM Salar had over-extracted water and brine. In a 2018 Commission of Inquiry into the agreement between CORFO and the company, it found that SQM had breached the terms of its environmental conditions of extraction, had over-exploited areas of the Salar, and had modified the variables in its environmental mitigation plans without environmental authorization. After the company was fined and denounced in 2019, the SMA agreed to the company's plans to spend 25 million USD to comply with the law,

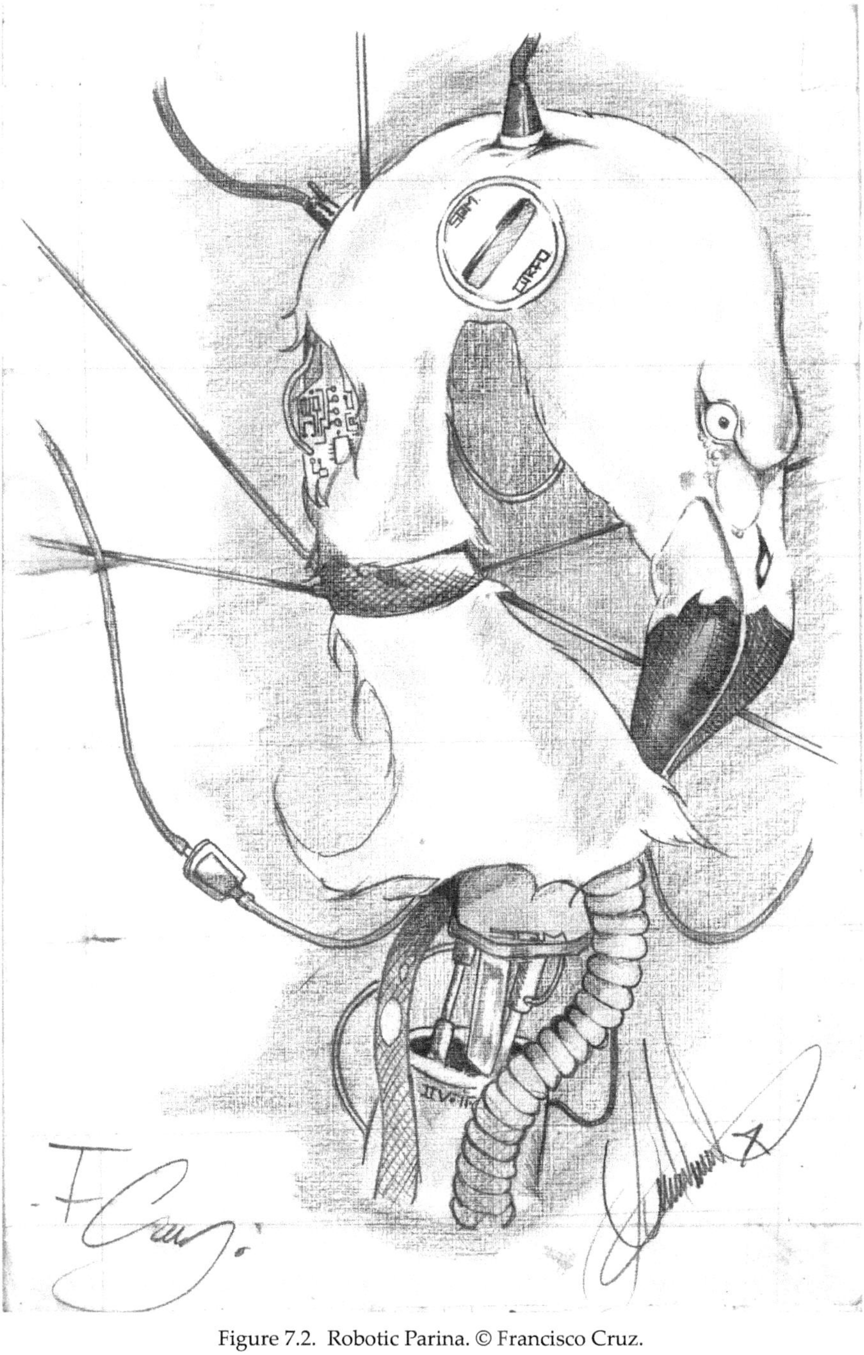

Figure 7.2. Robotic Parina. © Francisco Cruz.

plans which the regional Antofagasta court then found were 'insufficient', which in turn meant the case went to the Supreme Court. Later that year, the Indigenous communities of Peine and Camar and the CPA filed a claim with the First Environmental Tribunal protesting the environment department's approval of SQM Salar's mine and plant expansion on the Salar de Atacama on the grounds that the company did not consult with communities, as required under the terms of C169, and that the environmental impact mitigation plans were inadequate. The Tribunal found in favour of community claims.

Despite legal findings against SQM the company has continued operating, and has in fact grown. SQM's public declarations embed assumptions about the amplification of extractive activities until 2030, after which under President Boric's Lithium Strategy, CODELCO, Chile's national copper corporation, will take on an operating share. Peine, the CPA and a number of other Atacameño communities of the Salar have taken the company and CORFO to court, as well as entered into negotiations with SQM and the state body CORFO, and since 2023, some discussions with CODELCO. At the time of writing, these negotiations were interrupted, as four of the Salar de Atacama communities (Peine, Socaire, Toconao and Camar) released a declaration suspending their participation due to concerns about the unequal effects of lithium mining and the lack of recognition of these inequalities in the ongoing dialogue.

Given this book's discussion of the imagined and actual use of agreements by Indigenous peoples, and despite their risks, such negotiations and their troubled entanglements would come as no surprise. However, the political economic relations between SQM and first CORFO, then CODELCO, which have become partnerships of public and private interests, have begun to drive the extractivist agenda into the 'critical minerals' future. Refusal to accept the terms of this kind of powerful alliance is complex, though sporadic community protest actions have more consistently marked the regional political landscape. Atacameño community leaders and members have also responded to the conditions of participation in SQM's mining future by troubling its very terms and by seeking to engage on terms that better suit them.

We now turn away from the nucleus of the Salar back to its southern fringes, to the foothills, canyons and mountains that lead down to and overlook it. As extractive excesses expand, and the slow resistance of Atacameño peoples becomes part of national and international public attention, it may be easy to forget that at the same time communities are busy engaging with the ongoing extractive pressures from the copper industry. Alongside the lithium boom, there are many other sets of engagements played out between the Peine community and extractive industries in the context of the new environmental regulations. Two further examples of consultation under the new environmental regulations depict the scale of work, the overburden, that Peine's leaders diagnosed, and the ways in which participatory mechanisms of extractive governance are part of but also trouble advanced extractivism. The first is a brief example of a proposed new copper mine, and the second returns us to the case of Peine's relations with Minera Escondida.

A Dolphin in the Foothills

In 2019, Joaquin, speaking as the President of Peine told journalists:

> The Minister says that the authorities are open to listen to the companies that want to develop sustainable projects. If this project was sustainable, it would have been approved in 2016, when the Regional Commission of Antofagasta first voted, [but they] unanimously rejected the proposal. Mysteriously, the Council of Ministers has given a second opportunity for the company to put its proposal in front of the SEA. This is really bad faith on the part of the state. (Cardenas 2019)

Joaquin was referring to a mining project called *Minera Delfin* ('Dolphin'), whose owners proposed to excavate a copper mine seven kilometres from the locality of Peine. The project proposed to use technology of underground blasting rather than extract by creating a large open pit and to cart the product away rather than process the mineral in situ, both material dimensions of the project that the proponents claimed would produce only low environmental and social impacts. From 2010 onward, I had the opportunity to observe some five meetings between the owners of Minera Delfin and community leaders as the company sought to minimize community resistance to the project. I was later told by leaders that proponents had ushered them aside to give them gifts, and in meetings I observed had made offers to 'help out' if leaders agreed to a deal. The proponent's 'little project' (reportedly worth only 17 million USD), was first described to the SEA in an impact declaration (DIA). The DIA was rejected by the regulator in 2011, and the proponent was forced to go through a full Environmental Impact Assessment process including Indigenous community consultation.

Minera Delfin submitted their EIA report in early 2013. The timing of the reporting and consultation period intersected with community elections lasting some months, as well as ongoing community negotiations with Rockwood-Albemarle that took up the time and effort of the *directiva*. As a member of the community explained, 'the internal dynamics of the community often requires time in ways that don't fit with the schedule established by the public service' (Comisión de Evaluación 2019). In the tail-end of 2015, community members worked with a team of external consultants to devise a process of consultation about the Minera Delfin EIA in three stages; broadly, information provision, community analysis through consensus and internal deliberation, and finally, dialogue and outcome. Minera Delfin proponents and the government agency provided written and verbal information about the proposed project to an *asamblea,* and around forty members of the community undertook a fieldtrip to the site of the proposed mine and deliberated the project in a further special *asamblea.* The consultation team engaged in door-to-door discussions and ran focus groups in the village according to generation and gender and then discussed the project in two more special *asamblea.* In a final meeting, the *asamblea* voted to reject the project outright and to call a halt to engagement with Minera Delfin. Community leaders travelled to Antofagasta to report their decision to the Commission of Environmental Evaluation. The regional body upheld the community decision, ruling that the proposed project could not go ahead.

Minera Delfín appealed. A year later around a dozen members of the Peine community travelled to Santiago to stage a protest outside the government building, while the Commissioners inside deliberated. In Peine, around 100 people marched along the *camino tropero,* the path used by herders for hundreds of years, leading to Tilomonte, which would be severed if the proposed mine were to go ahead. Marchers held banners and placards, with their chants and *cacerolazo* resounding out across the Salar.[11] In Santiago, the Commissioners ignored widespread community rejection and found that *consultation* was deficient, ordering the proponent to reengage with the community. Community leaders released a statement that they would 'not dialogue with the owner, nor are we in agreement with the project'. By 2018, when the proponent again sought to prepare an EIA, it engaged two consultants to speak with community members about the social impacts of the project. The consultants were met by a village of houses flying black flags and closed doors.

In December 2018, the *directiva* asked several researchers (and I) to assist in compiling a critical analysis of Minera Delfin's EIA.[12] My assessment of the section on social impacts was that the errors were extensive, though largely because the consultants had not had access to locals. The report failed to address previously articulated local concerns, including the effects of blasting on underground water in the area of the proposed mine, the socioecological impacts of runoff of waste from the mine to pasture areas on the border of the Salar, the impacts of mine workers and hauler truck noise, and the potential effect on the largely unregistered cultural heritage values of the area.

Community leaders were incensed at how the state had not taken their participation seriously. The company had been allowed to continue to lobby the Minister of Environment despite the fact that the Regional Council of Antofagasta had reviewed the original report and rejected it unanimously. Eventually, the new copper mine Minera Delfin was defeated and most community members expressed relief that it would not go ahead.[13] Nonetheless, the many years of dealing with reams of bureaucratic procedure, conflicting technical advice, campaigns for local, regional and national government support, the pressure of local deliberations and managing fiercely divergent opinions all took their toll. The burdens of refusal and negotiation are one and the same in the sense that both entail significant resources to engage with the technical, moral and material politics of extractive interests. Peine's experience of refusing Minera Delfin's mine buoyed the sentiment of community leaders I spoke to about the possibility of being heard if they said 'no' to extractive interests that threatened their future. They also said that participation is exhausting, preventing them from doing all kinds of other things.

Late Capitalism's Too-late Regulation

The material politics of uncertainty and corporate science have overdetermined extractive harms in the southern Salar de Atacama. In the case of Minera Escondida, long-installed in Atacameño territories, the environmental regulations came too late for the Peine community to have any say over the installation of Escondida's extensive infrastructure, and over the harms that the company's groundwater extraction has produced.

As Table 7.1 shows, Escondida began extracting groundwater for copper processing in the 1980s from around the Salar de Punta Negra, which is located around 100 km south of the Salar de Atacama. Starting in the late 1980s, Escondida was permitted to purchase water extraction rights in the area totalling over 1,500 litres per second drawn from forty-eight wells, plus another 150 litres per second of surface water extraction points.

The Peine community consider the Salar de Punta Negra to be part of their territories. Flamingo eggs and feathers were harvested there, and the scant surface freshwater at its edges was available for pasturing animals. As a result of complaints from Peine Community members and others, in the year 2000, the Chilean Agriculture and Livestock Service (SAG) found that Escondida's water extraction had caused damage to the Salar de Punta Negra, as well as an impact on flamingo breeding grounds;

Table 7.1. Mining water concessions, Salar de Punta Negra (and surrounding basin).

Company	Location extraction	Quantity (litres/sec, annual average)	Year range
MINERA ESCONDIDA LIMITADA	Endorreicas Salar Atacama-Vertiente Pacifico. Pampa de Socompa x6	234.4	2001
MINERA ESCONDIDA LIMITADA	Endorreicas Salar Atacama-Vertiente Pacifico. Salar de Punta Negra x 69	1394.84	1986, 1987, 1990, 1992
MINERA UTAH DE CHILE INC.	Endorreicas Salar Atacama-Vertiente Pacifico. Salar de Punta Negra. Quebradas, Zorras, Salado, Llullaillaco x3	4	1985
COMPAÑIA MINERA ZALDIVAR SPA	Endorreicas Salar Atacama-Vertiente Pacifico. Salar de Punta Negra x1	8	2009
Total		1641.24	

Note: See Table 2.1 for a discussion of limitations on this data, as collated from the DGA register. Table 7.1 focuses on Escondida, but Zaldívar, operated by Antofagasta Minerals, submitted an EIA in June 2018 to continue their operations, including water extraction in the Negrillar sector, until 2031 (documents available at http://www.seia.sea.gob.cl). At the time of writing, that matter was unresolved.

however, it did little to halt further ecological exploitation. The company signed a 'good faith' agreement with SAG in which they would pump water back into the Salar de Punta Negra and undertake other flora and fauna mitigation programs to address the ecological harm caused (see also MEL 2006). People reported industrial waste and rubbish left in the area by Escondida and other extractors and noted that pumping water back in had not regenerated wetlands and flamingos were not breeding in the artificial nests (TVN 2007).

In 2016, with elected officials willing to begin using the revised environmental regulations, Escondida was under pressure to address concerns over the Salar de Punta Negra. Regional politicians highlighted the failures of the SAG and Escondida agreement, stating the environmental harm may be traced to lack of government oversight. Senator Navarro was quoted on regional television news channel Antofagasta TV as saying, 'when I asked [the Conservation and Forestry Department] CONAF how many flamingos are on the Salar, they told me to ask the company'. With this ongoing pressure, as well as impending negotiations over its Monturaqui EIA (see Chapter 2), the following year saw Escondida announce it would cease all further extraction from the Punta Negra basin. However, their activities had already done considerable harm. I return to this below.

When Escondida released their Monturaqui EIA, it first went through a period of citizen consultation. In these initial stages, Peine's leaders requested that I put together an analysis of the 'social impact' sections of the EIA for them.[14] I arrived in Peine in August and met Miguel in Dona Alma's store one morning. He asked me if I had read the EIA yet. I told him I had begun, but its *table of contents* was 58 pages long, so it was taking me some time. The report came in three volumes, was over 2,000 pages and contained reams of hydrological data, with a 172-page-long section on social impact ('Medio Humano'). Miguel scoffed that it was a waste of his time, but he had read the important parts. What annoyed him most, he said, was the company's claim to be undertaking 'voluntary' measures of mitigation and compensation (e.g. Figure 7.3). These measures included scientific studies to assess the MNT aquifer and its relationship to the surface ecology of the southern Salar. Given the fact of environmental regulations protecting Tilopozo, 'voluntary' could hardly describe a response to years of community pressure. Miguel's critique echoed others who, as they began to read the detail, found Escondida's claims historically, scientifically and legally problematic.[15]

Community leaders invited me to a meeting of the *directiva* to discuss the Monturaqui EIA. Joaquin, opened by providing a history of their relations with Escondida, beginning with the 1997 'good neighbour' agreement, and then the better deal in 2007. In recent years, he recounted, the community tried again to renegotiate the agreement as they became more concerned about Tilopozo, arguing that Escondida must reduce groundwater extraction and put in place greater measures of compensation. The company resisted, until it released the Monturaqui EIA when 'measures of compensation and voluntary promises' were included that reflected community demands. This history, Joaquin said, showed some of the problems in trying to deal with Escondida. Like Miguel, Joaquin's account of the labour of their relationship and of community demands refutes the company's account of voluntarism. Far from 'offering' compensation, Escondida had been under substantial pressure to take reasonable steps to mitigate extractive harm.

For its part, Escondida had repeatedly requested as part of these discussions that the community agree to 'social baseline' research by the company's social performance team. For example, in 2011, Ana and Nicolás from Escondida's Community relations team raised the proposal at an *asamblea* to undertake a survey that would provide the community

Figure 7.3. Pamphlet produced by Minera Escondida to summarise the company's 'voluntary' and compensatory offers to the Peine Community at the time of consultation processes associated with the 2017 Monturaqui EIA (reproduced from copy of pamphlet held by the author)

and the company with a 'baseline study'. My fieldnotes reflect that the *asamblea* met Escondida's request with silence, though after Escondida employees left the meeting a handful of Community members raised the possibility of compiling their own study which they might wield in their own way. Baseline studies provide a quantifiable beginning from which social and environmental change may be measured or more confidently indicated; a project ground zero for impact. The problems with this instrument, especially as it is embedded in the EIA process, are manifold (see Tecklin et al. 2011; Ureta 2018; Barandiarán 2019a). There were no such studies undertaken in the Salar prior to the beginning of mining. While any point chosen as the 'baseline' from which 'impact' is counted is arbitrary, it is especially problematic to do so for just one company. With the number of extractive projects in the area, any effective indicators of social or environmental change would need to consider their sum. As critical commentators have argued (Ureta et al. 2020), baseline studies are both limited and useful exactly because they are politically malleable.

Community leaders were aware of the use of data for company purposes. Upon embarking on an analysis of the Monturaqui EIA, they gave me specific instructions to focus on community concerns about groundwater and not to reveal contemporary socioeconomic or cultural information. This was part of a long-running effort to emphasize contemporary politics and community concerns for the future rather than reveal social, cultural and economic detail of 'community business' in a baseline study. Nonetheless, Miguel told me he felt their refusal to do a baseline study had been a risky strategy, which '*nos costó harto*' (cost us a lot), given that company employees' manipulation of leaders, partly in regard to this matter, had caused local disputes. The burden of refusal included social tensions that were still felt many years later.

A refusal to open their community to a baseline study was informed by knowledge of the company accumulated over years of endurance of the relationship, including the internal trouble caused by community relations employees from Escondida (see Chapter 6), the history of the company's environmental damage in the Salar de Punta Negra and fears of this being the fate of other waters such as Tilopozo. Moreover, refusal was fuelled by the mounting extractive burdens in the southern Salar. Community members repeatedly asserted they were unconvinced by company science, and they were tired of unfulfilled 'community development' project promises. The twenty years it had taken for Escondida to build the Llullaillaco stadium in Peine stood testament to the community's endurance of such relations. It was the myriad of such burdens shouldered by the community and their hope for life and a future with watered places that, when it came to the vote in January 2020, led to an overwhelming 'no' to Escondida's proposition to continue to extract from Monturaqui. This time they had regional politicians and improved regulation on their side. A news item covering Escondida's retraction of their EIA and announcement that the company would no longer extract water from Monturaqui and the company shut off the pumps.[16] It quoted Esteban Velasquez, member of Antofagasta regional government celebrating Escondida's announcement to halt groundwater extraction and condemning that the region was treated as a 'zone of sacrifice'.[17] The language of these accusations is important in the growing national critique of extractivism in the context of the boom for critical minerals in Chile, and especially the Antofagasta region.[18]

In April 2020, the Peine Community and the Council for the Defence of the State (CDE), brought a case to the First Environmental Tribunal accusing the company of maintaining extraction of freshwater between 1990 and 2017 from 'no less than five different projects', causing 'continuous, cumulative, permanent and irreparable' environmental damage to plant, bird, mammal and reptile life, as well as microcrustaceans, nematodes, insects and algae in the Salar de Punta Negra (Primer Tribunal Ambiental 2020). The Salar, these reports accused, 'will not be able to recuperate on its own over 100 years'. This is licit ecological damage. Escondida's activities received legal environmental permit, but none of these permits have included studies of impact on the hydrological system or outlined the company's measures of mitigation or compensation associated with that impact.[19] The material properties of the Salar de Punta Negra remote from the centres of bureaucratic agency, and also remote from any towns and permanent settlement, are significant in the capacity of a materially absent state actively enabling harm and the conditions of the unseen in this case relate to processes by which water's relations to the future is 'counted'.

In May 2022, a little over two years after Escondida had switched off its pumps, the Chilean Superintendency of the Environment (SMA), declared that the company's groundwater pumping in the Monturaqui area had caused environmental damage in Tilopozo. The harm Escondida caused was deemed by the SMA to be 'irreparable' and the agency levelled a fine of around 65,000 USD.[20] The company Escondida appealed. Before they shut off the pumps in February 2020, Escondida's environmental approval for groundwater extraction had relied on the company scientist-defined threshold in the aquifer as 25 cm (see Chapter 2). Up to this point, they had asserted, the water was sufficient to support surface vegetation systems in the wetland pasture at Tilopozo. Escondida's lawyers argued that the lithium miner Albemarle operates directly in the Tilopozo wetland and Zaldívar also extracts from the MNT aquifer. Neither were subject to these government sanctions. Following the SMA finding, the Peine community sued all three mining companies in the Environment Tribunal for having drawn on the MNT aquifer to the extent that harm would be done to the surface ecology. They were joined in the case by the CPA and Chile's State Defence Council (CDE) (Case D-12-2022). In part the legal case states the basis of the charge as –

> Tilopozo, constituting an oasis in the desert, plays an essential role in the economic and cultural activities of the Peine Indigenous Community, historically and presently constituting a gathering point for community members and their pastured animals, giving rise to the expressions of culture central to the role these ecosystems occupy in the Likanantay *cosmovision*.[21]

In response, Escondida's lawyers asserted that they had always conformed to environmental permissions and that: 'there can be no environmental damage if what is produced is an evaluated and approved impact'.[22] The statement is a neat summation of advanced extractivist conditions by which ecological harms occur, both below and above the surface, in a process that is entirely legal. (At the time of writing, the case is not yet resolved).

The EIA is a regulatory tool for decision making on industrial processes. As Morgan (2012) argues, it identifies as a theory of rational choice in regard to 'best solutions' to the technical problems of development and impact and seeks to 'bring community members into the process' earlier in 'development' processes (p. 8). However, the centre of gravity of a definition of 'impact' is with the powerful, who define, mitigate and consult about it. While the EIA as participatory mechanism has facilitated greater inclusion of Indigenous people – communities have gained 'a seat at the table' after many years of demanding one – the terms by which they may come to that table, and the implications for the 'choices' made remain structurally uneven. Communities like Peine may be able to articulate responses to the environmental problems of multiple projects at once, but the burden of struggle and engagement itself has significant social and ecological implications.

Conclusion

In the Salar, the conditions of the lithium boom have catapulted the apparently mundane and largely invisible effects of extractive industries into a globally recognizable landscape of mining conflict. As lithium mines have taken up ever greater areas on the surface of the Salar, in full view of the eastern Salar's villages and residents, the area more and more has reflected the stereotypical characteristics of an environmental crisis and fit the critique that it is part of the broader region as a 'zone of sacrifice' to the wealth of the nation. Global media attention on the ecological impacts of the lithium boom also attracted researchers seeking to speak with community members, and these researchers have focussed on the story of people suffering the impacts of global capital and extractivism. This rising crisis, a spectacular series of events including protest and legal controversies (only some of which I have detailed) belie the hidden dynamics of the extractive industry prior to the lithium boom. They also overlook the mundane dimensions of crisis. The social and ecological effects of the lithium boom continue to occur below the surface. The community's critical complaint regarding 'overburden' alerts us to these relatively obscured and structural conditions of advanced extractivism.

The moral virtue of Indigenous rights to territory and to self-determination is linked to an imaginary of Indigenous peoples as collectivist ecological stewards, opposed to rapacious capital in the politics of extractivist conflicts. Troubling such visions, as Indigenous peoples have done elsewhere on the planet, Atacameño peoples participate in regional, local and extractive governance of the Salar de Atacama through measures associated with C169 and with environmental regulations, and also conform to the image, engaging in overt resistance and political activism against extractive interests. These varied actions have material effects, demonstrating both the political value of the stereotype of 'environmental steward' and its limits, pushing extractivist logics by expressing concern and outrage in the face of harms while also engaging with mining company actors in ways that appear to parallel a rhetoric about 'responsible business'.

Atacameño people have gained material and political ground through such tactics. They have responded actively to the introduction of government regulations regarding obligatory reporting and community consultation over environmental impact of new and planned industrial activity. As they have done for many years, communities like Peine leverage mining company 'development' funding that flows from negotiations over impending or actual extractive activity. In the Salar, Atacameño leaders have also been able to travel to meetings like COP25 in Madrid and to the Chilean capital to protest against extractive excesses, and also to meet with other Chilean Indigenous leaders, to build their professional teams and pay lawyers, and to lobby politicians about ecological and social impact. Using multiple tactics in the range of negotiation and resistance also carries a range of social and moral burdens.

The temporalities of *asamblea,* as both event and polity, potentially provide a measure of community control over the radically unequal conditions of participation in extractive governance. Amidst pressure from outside interests, 'the *asamblea* decides' can be used to alter others' temporalities, a powerful strategy to slow down decisions. This comes with its own burdens. In the case of Minera Delfin, the Peine community troubled the process by slowing consultation, then refusing to participate once they had community consensus to do so. They were able to stop the new copper mine from breaking ground. Again, Peine have continued to draw 'development benefits' from Escondida while also taking the company to court, and slowed future groundwater extractions by blocking proposals, most recently the Monturaqui project. The Environmental Tribunal is a new resource for this slow resistance, but may be too late to prevent ecological harm to Tilopozo and the Salar de Punta Negra. And the environmental regulations, which introduced mandatory consultation of Indigenous peoples in the same period that the lithium boom was emerging, has burdened communities with massive and competing demands.

As projects have mounted, how might exhaustion be prevented? Not only must Indigenous communities find the resources to deal with technical problems of company science, but they must come to community decisions (through collective deliberation) about the problematic details of every mining or industrial project. Because one of the aspects of the overburden of consultation is internal social conflict, it is largely unseen, or patently ignored, and rolls consistently forward with the demands of consultation in the environmental regulation process. Peine's decision not to have a baseline study, for example, rejected the idea implicit to such reports that an arbitrary 'ground zero', from which ecological relations change through 'impact', could be defined by the mining company. The decision emphasized their authority over ecological and human relations of the Salar and broader territories. Moreover, their refusal to consult with proponents in a way proponents find convenient is a way of resisting the exhausting inequalities of ecological and social effects produced by the extractive industry.

A mundane form of political and community activity in the Salar intensified in response to progressive state legislative and regulatory regimes. Extractive effects for Atacameño peoples always come with concerns for the ecological future of the Salar de Atacama, but they are also about the everyday problem of administrative and technological details of the paperwork about extractive excesses. Chile's neoliberal state failed to provide scientifically informed environmental governance capacity, and progressive political change has also been committed to the lithium industry in the Salar de Atacama and its growth. State agencies now oversee many dimensions of extractive activities, and environmental tribunals have taken on important cases of extractive harm, but the ecological future of the Salar is far from ensured. For Atacameño peoples, as for other Indigenous communities dealing with similar forms of extractive participation and governance, an excessive burden of evaluating the scientific claims to sustainability in reports by extractive capital falls to them.

Notes

1. The conference is organized by a research institute situated in my employer university, the Sustainable Minerals Institute at the University of Queensland. It organizes a biennial conference on Social Responsibility in collaboration with transnational mining companies. SMI organizers agreed to waive my conference fees and in return I chaired a bi-lingual session of the proceedings.
2. The 2016 Chilean senate inquiry into glaciers and salt pans and media attention associated highlighted the threat in the national imaginary. In the subsequent citations I have selected from media reports in the European and North American press; however, it should be noted that some have significant errors of fact. The rush of attention is the point I am making.
3. For example, bilingual article by Centre for Interdisciplinary Environmental Justice (2019). See also Riofrancos (2019) and the Business and Human Rights Resource Centre (an NGO based in the UK) wrote an overview report that garnered response from Albemarle (http://www.business-humanrights.org/en/chile-concerns-around-social-environmental-impacts-over-rockwood-lithium-operations-acquired-by-albemarle).
4. The map, which allows you shift the view 'before' and 'after' lithium mining (and which includes in the prior image, taken in 1993, the Sociedad Chilena de Litio, now owned by Albemarle), is available online at: http://www.usgs.gov/media/before-after/lithium-mining-salar-de-atacama-chile. When I wrote to the office asking why they featured the Salar in 2019, they replied, explaining that it was simply that the office 'routinely posts images of interesting events or changes that are taking place on the [Earth's] surface'. (email correspondence ticket #274574, US Geological Survey Science Information Services, 5 June 2020). The 'after' image is out of date, with major expansions in both plants since 2015.
5. These refer to the same mine. When Peine negotiated an agreement with the lithium miner it was still operating as 'Rockwood Lithium', as Albemarle took over operations of that mine their agreement with the CPA, to which Peine is a party, was signed in the name of Albermarle.
6. A 'declaration' can only be made in cases that do not have a certain level of environmental and social impact. In Rockwood/Albermarle's case, the company was able to undertake minimal consultation for its environmental license for two reasons; first because their expansion plans had been lodged in 2009, just before the new legislation came into effect, and second because the company signalled their readiness to come to an agreement for 'shared benefits' with the Peine community (thereby apparently satisfying the need for consultation). I discuss SQM below. In both Albemarle and SQM's terms, the location of the plants, evaporation ponds and groundwater extraction in the Salar de Atacama was on state land (CORFO) that had never been recognized as Indigenous territory; therefore, community consultation was considered legally unnecessary.

7. Chile's hosting of the COP25 meetings was cancelled because of the radical unpredictability of the *Estadillo*, the ongoing street mobilizations in the capital and throughout the country.
8. This is a partial transcription from online video of the event (published on the CPA Facebook page). Referring to 'the Salares' in plural, supposedly including the Salar de Maricunga, and 13 other saltpans in Chile rich in lithium (see Cabello 2022).
9. As I have attributed the drawing to him, this is not a pseudonym.
10. According to Barandiarán (2019b), the government's renegotiation of Albemarle's contract to extract lithium included higher royalties to be paid to the state, as well as the requirement that the company enter into 'benefit sharing agreement with local communities' (p. 387). As I've discussed, the signing of agreements between Rockwood/Albemarle and Indigenous Communities were driven by Community insistence and activism around these issues.
11. The *cacerolazo*, banging pots and pans. The march was possible precisely because the proposed mine site was so close to Peine. Leaders began to take advantage of social media to broadcast local opinion, liberally using the social media platform Facebook to link their remote 'voice' with news media and environmentalist support.
12. I undertook the work 'pro bono' and remotely; I compiled a report online and via telephone call.
13. There was always a small number of Peine Community members who were in favour of the Minera Delfin proposal for their reading of its economic promise.
14. A period of fieldwork during previously planned time in Santiago from Australia enabled this to be done in collaboration and without research cost, though the Community funded one additional (otherwise unplanned) trip from Santiago to Calama and accommodation in Peine's Guest House.
15. A hydrogeologist was engaged by the Community at the same time as they asked me to write an analysis of the data problems in the EIA and its accompanying assertions. Our reports were incorporated into the Peine Community's response to the citizen consultation process that precedes the Indigenous consultation stage.
16. Antofagasta TV reported (on 5 February 2020). Available as 'Pressure from communities permitted the non-extension of use of water in Monturaqui' Antofagasta.tv/2020/02/05/presion-de-las-comunidades-permitio-no-extender-uso-de-agua-en-monturaqui-por-parte-de-escondida/ (last accessed 3 March 2020).
17. In Chile, a critical and popular language on 'sacrifice zones' has developed in relation to areas in which industrial pollution and environmental degradation are intractable or irreversible problems of neoliberalism, unable to be solved, since they are a function of abandonment, by the state or government institutions. Consider, for example, feminist and ecological work (Bolados García and Sánchez Cuevas 2017), a collection of critical essays (Programa de Reducción de Riesgos y Desastres 2022) and a work of narrative journalism (Contardo 2022).
18. For example, the Governor of Antofagasta (*Intendente*), appointed by the central government (in Santiago), made public statements critical of the regional government's apparent support of community concerns in regard to Escondida's environmental impacts. In response, the regionally elected politician commented that 'the irresponsible Chilean state would rather sacrifice our region in order to maintain the businesses that support centralism' (Regionalista.cl online column, 23 August 2018).
19. This notice is available as 'First Environmental Tribunal receives legal suit for environmental damage against Minera Escondida', Primer Tribunal Ambiental webpage (www2

.1ta.cl/primer-tribunal-ambiental-acoge-a-tramite-demanda-por-dano-ambiental-en-contra-de-minera-escondida/, last accessed 16 April 2020).

20. Considering Escondida mine uses more water in this location (to wash and process copper ore) than any of BHP's other mega mines in other parts of the world (see BHP 2019), and the size of the Escondida mine and its profits, the fine was very small.
21. Exact quote. '*Tilopozo, al constituir un oasis en medio del desierto, cumplen un rol esencial para la actividad económica y cultural de la Comunidad Indígena Atacameña de Peine, constituyéndose histórica y actualmente como un punto de encuentro de los comuneros y sus animales de pastoreo, lo que da lugar a manifestaciones culturales derivadas del rol central que estos ecosistemas ocupan en la cosmovisión lickanantay.*'
22. Exact quote: '*no puede haber daño ambiental si lo que se produjo es un impacto evaluado y aprobado*'.

Conclusion

A combination of the extreme surface aridity that seems to signal limited ecological capacity to support life, the remoteness from the centres of power and the small and racialized population of the desert highlands have enabled the Salar de Atacama to be slowly engulfed by the surrounding 'mining region', valued in dominant terms for resource wealth. Progressive governments have incrementally introduced environmental regulation over the past thirty years as the industry has grown, and a sense of ecological crisis in the southern Salar de Atacama has surfaced. Critics say that the Salar is one *more* area of Chile that the state has 'sacrificed' or 'abandoned', along with its peoples and ecologies, in pursuit of political economic wealth and, in this case, to meet the global desire for lithium (e.g. Jerez et al. 2021; Azócar 2022). This book's focus on the recent history of extractivism, largely from the late twentieth century to the present, reveals how the contemporary mining landscape – in which ecological and social effects are becoming more visible, more widely contemplated, and seem more certain – can be traced to the apparent uneventfulness of extractivism in the Salar de Atacamain the decades before. In the Salar, the material and geophysical dimensions make it distinctive for understanding advanced extractivism as a globalizing and also very particular planetary dynamic. The moral thrust of extractive development, that is, the assertions that extraction may be 'good for communities', is intimately connected to ecological and political relations over time that align with, as well as trouble those claims. In summarizing some of the book's arguments, this brief conclusion combines different flows and tensions of groundwater politics.

* * *

When we spoke in 2009 in San Pedro de Atacama, Doña Gracia told me 'the Indigenous Act was not what we had hoped'. Throughout my research leaders like Gracia helped me understand the implications of late twentieth century Chilean state recognition of Atacameño peoples and their rights. The process was limited, but it potentiated some social and political change driven by Indigenous Peoples. She told me,

> It [the Act] was something, but we had hoped for more. We had seen the draft of the Indigenous Act and made some improvements, but we realised that we would have to do something profound since a giant was upon us, tourism and mining and many other things, and we would have to protect it [water] in some way. This protection is the only way because with the title owned by the pueblo, that is all the people, then to sell water all the people must come to agreement and that would never be the case. It would never be the case. Precisely for this reason, we campaigned in the Salar [among Atacameño people] to register all the water as communal title. The best thing we achieved was to [do this and therefore] protect our waters. It meant that the water was inscribed for all the people.

The potential for registration of individual titles to water threatened to divide Atacameño territory, water, and people, as it continues to do. With Doña Gracia and other Atacameño leaders' prescience, people campaigned for collective title and communities registered as collective rights the scant but critical superficial waters that nourish the needs of the living around the Salar. While far from total protection of waters of the area, their actions were a statement of pre-eminence of an Atacameño value for water as future life (see also Mora and Rivera 2017). Responses like Gracia's to my questions directed me to see water in the Salar de Atacama as the very substance of politics, nourishing all relations among extractivist and Indigenous ecologies and humans, their conflicts and divisions, as well as agreements and intimacies.

More than a decade later in Peine, when I was discussing the final draft of this book with Dolores, she told me that she didn't like the term 'slow resistance', saying 'no, not resistance. *The Chilean state* made mining possible'. Her implication seemed to be that people in Peine could not be represented as 'resisting' since mining was enabled by the government, not by them. Indeed, my desire to use the term 'slow resistance', might seem out of place for others (in addition to Dolores), whose familiarity with the Salar and northern desert regions of Chile is more intimate than my own. I retain the term as a provocation to consider that resistance may be a refusal that bubbles below the surface of events. It asks the reader to be alert to people's insitence on remaining themsleves and characterizes practice like collective work and registering community titles to water as resistance to dominant expectations of individual property rights that remain part of Chile's extreme version of neoliberal society and political economy.

Slow resistance indexes the many ways that people respond to the 'giant upon' them and their territories, and their recurrent refusals that the giant(s) wholly define their future. This is slow resistance, just as the state's overallocation to lithium and copper mining companies of groundwater in the Salar, which is gradually and incrementally exhausting ecological lives, is slow violence. The effect of Atacameño collective actions is an expression of their alternatives in a material politics. It is a refusal of ecology and economy conceived simply as material not also relations put to use for human gain in the radically uneven contest among diverse actors in extractive landscapes.

Advanced extractivism presents a moral, political, economic and technical problem, at once enabling, empowering, and incorporating people into industry and subjecting people and ecologies to exploitation. In northern Chile, as elsewhere, the mining industry is part of the history that defines what it is to be human, in everyday life, and is embedded in questions of the aspirant, as well as distant future. The relation of people and mining, in Strathern's (2020) terms, is an interiority. These are regions of nations in which school principals may profess pride in graduating students, 'who . . . thanks to their own efforts and our teams of teachers, all received a quality education, based in the values of the mining sector' (*Estrella de Iquique* newspaper 29 December 2023, p. 3). Northern Chile and the migrants, workers, and Indigenous peoples are miners as well as intimate critics of mining's worst effects. In advanced extractivism, this is increasingly obvious; the industry and Indigenous peoples and communities are entailed in one another's material interests, while the relation is contingent on factors themselves interior to the relation.

The politics of groundwater in relations between the extractive industries and Indigenous Peoples is material; territories hold resources, and the same territories and waters hold histories, lives and imagined futures. As such, the notable absence of any mention of water in the contractual terms of the first and second agreements between the community of Peine and the mining company Escondida is an important omission. In the conditions of advanced extractivism, Escondida has dominated vertical technologies, sustaining a data, resource and property relation to the earths and waters they extract and impact. This relation limits material relations to groundwater for all others, while drawing communities and the state into promises about development benefits of ethical capitalism. Nonetheless, the absence of the matter of water and territory in company-community agreement documents, as well as increasing awareness of uncertainty in scientific knowledge of surface and subsurface ecology, unfolds the material politics among miners, Indigenous Peoples and groundwater. The absences and uncertainties reveal how few interests in common are held by communities and companies who otherwise may be parties to an agreement.

In considering how people in Peine relate to the waters of their territories, I have shown how human labour of different kinds generates water's value, rights and ecological history, and how human labour generates and regenerates relations among people and other beings (such as water) that inhabit the world. Labour maintains a sense of mutually constitutive being, though one never knows what may happen given the contingencies of human agency and the uncertainties of action by other living beings. Thinking in terms of material relations, both absent and present forms, is a way of emphasizing how uncertainty might be a useful principle in a strategy for ecological care. It may return us to the ethical question 'is this a relation to water that generates enduring and future water relations?'

Mining company-Indigenous community agreements are criticized by environmental activists and critics of the industry. As I outline in detail in Chapters 4 and 6, the agreement is also critiqued by scholars as a globalizing instrument of late capitalism that subjects Indigenous peoples to the material values of extractivism. In the first criticism, Indigenous people are accused of 'selling out' their territory, water or ecological values. I have heard people who work in the area of Indigenous rights, as well as those who seek to do the ethical business of mining companies say that 'one of the problems of the Indigenous consultation process is that it encourages Indigenous communities to be *rentistas*' or, 'rent-seeking', using the Indigenous estate to seek industry funding. In the second version of the critique, Indigenous people are represented as lacking in agency, imagined as without the political vision or ability to enact projects of their own devising. Both ideological positions limit the potential for alliances between Indigenous peoples and those who seek to minimize ecological harm from the extractive industry, and assume that if Indigenous people have agency, this political will should align to a set of liberal values from which the philosophy of recognition too arises. This book's analysis of the relations between Atacameño-Likanantay peoples in Peine and the various extractive companies operating in their territory has elaborated a different kind of critique, one that has come from dialogue with people in Peine about how they anlayse the conditions of advanced extractivism, see themselves within and experience these relations.

On the face of it, a critique of Indigenous tactics to engage and to make agreements with extractive industries embeds an assumption that Indigenous peoples have a choice; that mining companies acquired Indigenous consent to exploit Indigenous territory and use water for extractive gain. However, the idea of gaining Indigenous consent or engaging in any form of consultation is relatively new, with Indigenous rights and consultation processes arising in the twenty-first century as part of what I term advanced extractivism. All the mines currently operating in Peine's Indigenous territory broke ground before companies were required by law to undertake consultation of any kind; consent was not sought from community members, nor was it given. The fact of the material imbrication of company and community interests, the overlap of ecological and economic facts, was already and anyway established. As I have shown, these interests have been incrementally drawn together while simultaneously remaining disjunctive. In the case of the copper miners, their open cut mines are far from view, and harm from their groundwater extraction largely invisible and difficult to contest for all of the reasons set out in Chapter 2. In the case of the lithium mines, these are on the far side of the Salar in territories that the state has excised from Indigenous claims. As I show in Chapter 5, the first lithium mine began as a small-scale operation, offering good work to locals, productivity and a sense that life in place could be better with some adjustments to the new regimes of labour. These adjustments are perhaps a kind of pragmatism. However, as I show in Chapter 6, the first lithium mine was also deeply reliant on the surface and groundwater in Peine's territory, on the labouring bodies of the people, and on a careful sense of good relations. An analytic of material politics seeks to draw attention to the limits of ideological positions and to the ongoing critical Indigenous engagement with and refusals of relations with powerful outsiders.

Critics of engagements between companies and communities may also be positioned in the industry, as I show by quoting the SQM manager's cynical comments at the beginning of Chapter 6. His assumption that an agreement is community acquiescence to the exploitation of territory and waters in return for financial benefit misrecognizes how agreements may be more labile things. The Peine Community signed agreements focused on community acquisition of 'development benefits' for the village, and people have sought to finance local infrastructure projects, fund legal advice and professionals in the area of environmental protection, and fund improvements on ancestral homes. Since I spoke with the SQM manager of 'sustainability' the company was ordered by the Chilean state to distribute compensation for environmental damage from overexploitation of brines and to do so through making contractual agreements through CORFO with Atacameño-Likanantay communities. Catching a lift from Peine to San Pedro with CORFO employees one afternoon in 2019, I asked them what kinds of projects different communities in San Pedro were looking to fund with the SQM agreement money. In response, they lamented it was all about basic infrastructure provision such as electricity and potable water, improvements to houses and community hall construction. Communities were not proposing entrepreneurial projects or business ideas. Instead 'development benefits' were being sought first to provide basic infrastructure or modern facilities, filling the gap left by the state's lack of spending in Atacameño communities. In Chapter 3, the story of Chakisoke is an epic narrative about addressing the mundane need for freshwater; basic infrastructure. And it is worth noting that while the Peine community sought to use the funding from their agreement with Escondida for independent environmental studies of the southern Salar as a way of gaining advice about the veracity of company science, Escondida blocked the use of the funding for these purposes.

Chilean law has never recognized the full extent of Indigenous community lands and waters, nor Indigenous interests in capitalizing on assets where these are titled. This book demonstrates that advanced extractive technologies such as agreements are heavily weighted instruments that favour the terms of powerful corporations. In Peine's case, community leaders have actively sought agreements with mining companies in order to fulfil development desires of different kinds. Indigenous peoples' agency in agreement-making, both the instrument and the funding that flows from it, has been exercised to effect relations with powerful corporations, drawing them into contractual obligations of some kind, when few other benefits from the state or from companies are likely. My emphasis on community leaders' agency in agreement-making shows how such matters too are a material politics, rejects the 'ethical' company claims that company engagements with communities are ever voluntary, and details the ways in which agreements both fund community development projects and have, in the end, funded community monitoring of company activity in the absence of an effective state regulator.

Companies like Escondida have failed to secure community acquiescence to company water extraction despite years of attempted financial benefits distribution through agreements. The manager from the company SQM Salar I quote at the beginning of Chapter 6 might scoff, but Escondida's failure to secure Peine's agreement to its terms of continued extraction was not the unprincipled fact that the company had tried to 'trade' environmental harm for economic benefits. Rather, the always-present conviction among members of the Peine community regarding the importance of water and their ongoing refusal to comply with unwritten dimensions of these agreements allowed them to avoid Escondida's attempts to gain social license. Escondida failed to gain a 'social license to operate' despite making an agreement because at no time did the company manage to allay community fears that they were doing no environmental harm. Their operation, the most extensive groundwater extraction operation in the Salar de Atacama, was nonetheless sustained for many years by ineffective state regulation, groundwater's enigmatic relation to the surface, and company 'sustainability' reporting, all of which were consistently challenged by the community whose future is most imperilled by decades of excessive groundwater extraction.

Critique from analysts and occasional industry insiders, Deanna Kemp and John Owen (2017), has been that the 'social function' of the extractive industry is in a state of decline, and in relation to communities, 'mining companies are places of refusal and resistance, driven by industrial self-interest' (Owen and Kemp 2017: 492). They argue that 'countervailing power' is needed to 'reign in destructive tendencies' (ibid.), but that state and political interests, as well as non-state actors, are too bound up in the interests of corporations to be countervailing power. Instead, they argue in favour of internally regulating corporations and for a return to the 'social function' of the corporation, as the countervailing power within. This book details some of the dangers of assuming companies will regulate from within in ways that serve the interests of ecological futures.

The changes that have occurred in Chile, in which company 'voluntarism' in community engagement has shifted to a process that is somewhat mediated by the state, has seen the possibility for some support to community decision making. In particular, the creation of the Environment Tribunals have enabled Indigenous people to have their contest heard and allowed an avenue for professionals and the state to support their interests. Given the many uncertainties of ecological subjects like the Salar's groundwater, which is difficult to 'see', manage and monitor, and is left open to exploitation, how might bureaucratic, legislative and regulatory state agents and professionals consider the value of relational projects, rather than assume ideological positions?

The multiscalar dynamics of resource frontiers have perhaps never been more palpable than in the case of mineral extraction deemed critical in the 'global' energy transition, such as the case of lithium. Indigenous peoples' territories and everyday lives are inextricable from advanced extractivism and in the case of Atacameño-Likanantay peoples, advanced extractivism has generated a crisis centred on groundwater depletion and dispossession. It is becoming common knowledge that the effects are far from 'local'. While this book has centred on Indigenous worlds in the Salar de Atacama, these are not only illustrative of the geospatial or temporal point (the village, the boundaries of territory, or some set of cultural practices in the past), which may be conceived as *impacted by* the ravages of extractive industries. As other anthropologists have long argued, situating people within the dyad of 'global' to 'local' is a category mistake (e.g. Trouillot 2001; Tsing 2005), one that ignores how 'the local' defies 'spatial unity'

(Lambek 2011: 197–99) and positions 'the global' as that which intrudes in an imagined coherence of shared purpose and ethical life.

This book has sought to engage with and amplify Indigenous critiques of the globalizing practices of advanced extractivism. It has considered how, as with peoples faced with similar challenges, Indigenous peoples find themselves and their ecological relations subject to extraction processes of different kinds and at different scales. How they differentiate these matters, how they have engaged with the moral imperatives of extraction as development, and more recently, extraction for the 'global' energy transition reflects the profound inequities of these. Whose localities transition to imaginaries of the 'green' future? Lithium workers and community members negotiating with lithium mining companies, undertaking scientific projects that aim to counteract the limitations of mining company science, protesting against them and thereby speaking into globalizing processes, can generate powerful commentary about the inherent instability of futures tied to extractivism. Peine community members continue to forge relations with the mining industry in order to negotiate their future with few substantial protective mechanisms emanating from the state, under conditions of negligible state funding for 'development' (in some cases basic infrastructure) and with renewed emphasis from the state on extractivism. The transnational histories of Indigenous peoples, shifting territories, and the intimacies of emplaced realities for locals and non-locals (working locally) alike mean their critically local politics might better inform the emerging ecological politics challenging new rushes to extract earths and waters at any cost.

References

Acosta, Alberto. 2013. 'Extractivism and Neoextractivism: Two Sides of the Same Curse', in Miriam Lang and Dunia Mocrani (eds), *Beyond Development: Alternative Visions from Latin America*. Amsterdam: Transnational Institute, pp. 61–86.

Acuña, Roger Merino. 2015. 'The Politics of Extractive Governance: Indigenous Peoples and Socio-Environmental Conflicts', *The Extractive Industries and Society* 2(1): 85–92. https://doi.org/10.1016/j.exis.2014.11.007.

Alexander, William L. (ed.). 2009. *Lost in the Long Transition: Struggles for Social Justice in Neoliberal Chile*. Lanham, MD: Lexington.

Allen, Catherine. 2002. *The Hold Life Has: Coca and Cultural Identity in an Andean Community*, 2nd edn. Washington: Smithsonian Press.

Asociación de Academias de la Lengua Española. 2023. Diccionario de la lengua española, 'convidar', *Real Academia Española*. Retrieved 1 October 2021 from dle.rae.es.

Azócar Duarte, Rodrigo. 2022. 'La Instalación de la Minería de Litio en el Salar de Atacama y su Relación Inicial con las Comunidades Atacameñas. Obreros, Campamentos y Asistencialismo (Chile, 1962–1998)', *Estudios Atacameños* 68: 1–37. https://www.jstor.org/stable/48728325.

Babidge, Sally. 2011. 'Comparing Informal and Formal Community-Company Relations', *SRMining*. First Seminar on Social Responsibility in Mining, Santiago.

———. 2015. 'The Problem with "Transparency": Moral Contests and Ethical Possibilities in Mining Impact Reporting', *Focaal-Journal of Global and Historical Anthropology* 73: 70–83.

———. 2019. 'Sustaining Ignorance: The Uncertainties of Groundwater and its Extraction in the Salar de Atacama, northern Chile', *Journal of the Royal Anthropological Institute* 25(1): 83–102. https://doi.org/10.1111/1467-9655.12965.

Babidge, Sally, and Madeleine Belfrage. 2017. 'Failing Forward: A Case Study in Neoliberalism and Abandonment in Calama', *Cultural Dynamics* 29(4): 235–54. https://doi.org/10.1177/0921374017743300.

Babidge, Sally, et al. 2019. ''That's the Problem with that Lake, It Changes Sides': Mapping Extraction and Ecological Exhaustion in the Atacama', *Journal of Political Ecology* 26: 738–60.

Bainton, Nicholas A., and Emilia A. Skryzpek (eds). 2021. *The Absent Presence of the State in Large-Scale Resource Extraction Projects*. Canberra: ANU E-Press.

Bainton, Nicholas, and John R. Owen. 2019. 'Zones of Entanglement: Researching Mining Arenas in Melanesia and Beyond', *The Extractive Industries and Society* 6(3): 767–74. https://doi.org/10.1016/j.exis.2018.08.012.

Ballestero, Andrea. 2019a. 'The Anthropology of Water', *Annual Review of Anthropology* 48(1): 405–21. https:/doi.org/10.1146/annurev-anthro-102218-011428.

———. 2019b. *A Future History of Water*, Durham: Duke University Press.

Barandiarán, Javiera. 2018. *Science and Environment in Chile: The Politics of Expert Advice in a Neoliberal Democracy*. Cambridge, MA: The MIT Press.

———. 2019a. 'Documenting Rubble to Shift Baselines: Environmental Assessments and Damaged Glaciers in Chile', *Environment and Planning E: Nature and Space* 3(1): 58–75. https://doi.org/10.1177/2514848619873317.

———. 2019b. 'Lithium and Development Imaginaries in Chile, Argentina and Bolivia', *World Development* 113: 381–91. https://doi.org/10.1016/j.worlddev.2018.09.019.

Barandiarán, Javiera, and Sebastián Rubiano-Galvis. 2019. 'An Empirical Study of EIA Litigation Involving Energy Facilities in Chile and Colombia', *Environmental Impact Assessment Review* 79: 106311. https://doi.org/10.1016/j.eiar.2019.106311.

Barros, Alonso. 1997. 'Pachamama y Desarrollo: Paisajes Conflictivos en el Desierto de Atacama', *Estudios Atacameños* (13): 75–94. http://www.jstor.org/stable/25674662.

———. 2006. 'Reseña Historico-Juridica de la Territorialidad Atacameña (Siglos XVI-XXI): de Leyes y Titulaciones Interculturales', *Cuadernos Interculturales* 4(6): 9–35.

Barry, Andrew. 2013. *Material Politics: Disputes along the Pipeline*. Chichester, UK: Wiley & Sons.

Barthel, Thomas S. 1986. 'El Agua y el Festival de Primavera Entre los Atacameños', *Allpanchis* 18(28): 147–84.

Barton, Jonathon, Alvaro Román, and Arnt Fløysand. 2012. 'Resource Extraction and Local Justice in Chile: Conflicts Over the Commodification of Spaces and the Sustainable Development of Places', in Håvard Haarstad (ed.), *New Political Spaces in Natural Resource Governance*. New York: Palgrave Macmillan, pp. 107–28.

Bauer, Carl. 1997. 'Bringing Water Markets Down to Earth: The Political Economy of Water Rights in Chile, 1976–95', *World Development* 25(5): 639–56. https://doi.org/10.1016/S0305-750X(96)00128-3.

———. 1998. 'Slippery Property Rights: Multiple Water Use and the Neoliberal Model in Chile 1981–1995', *Natural Resources Journal* 1: 109–53. www.jstor.org/stable/24888446.

———. 2004. *Siren Song: Chilean Water Law as a Model for International Reform*. Washington DC: Resources for the Future.

———. 2013. 'The Experience of Water Markets and the Market Model in Chile', in Josefina Maestu (ed.), *Water Trading and Global Water Scarcity: International Experiences*. New York: Routledge, pp. 130–42.

Bear, Laura, et al. 2015. 'Gens: A Feminist Manifesto for the Study of Capitalism', *Theorizing the Contemporary, Fieldsites,* March 30. Retrieved 17 September 2024 from https://culanth.org/fieldsights/gens-a-feminist-manifesto-for-the-study-of-capitalism.

Bebbington, Anthony, et al. 2008. 'Contention and Ambiguity: Mining and the Possibilities of Development', *Development and Change* 39(6): 887–914. https://doi.org/10.1111/j.1467-7660.2008.00517.x.

Beckett, Greg. 2013. 'Thinking with Others: Savage Thoughts about Anthropology and the West', *Small Axe: A Caribbean Journal of Criticism* 17(3 (42)): 166–81. https://doi.org/10.1215/07990537-2378982.

Bennett, Vivienne, and Margreet Zwarteveen. 2005. 'The Connection Between Gender and Water Management', in S. Davila-Poblete and M. Nieves Rico V. Bennett (eds), *Opposing Currents: The Politics of Water and Gender in Latin America*. Pittsburgh: University of Pittsburgh Press, pp. 13–29.

Benson, Peter, and Stuart Kirsch. 2010. 'Capitalism and the Politics of Resignation', *Current Anthropology* 51(4): 459–86. https://doi.org/http://www.jstor.org/stable/10.1086/653091.

Beresford, Melissa, et al. 2023. 'Moral Economies for Water: A Framework for Analyzing Norms of Justice, Economic Behavior, and Social Enforcement in the Contexts of Water Inequality', *WIREs Water* 10(2): e1627. https://doi.org/10.1002/wat2.1627.

Bessire, Lucas. 2022. 'Aquifer Aporias: Toward a Comparative Anthropology of Groundwater Depletion', *Current Anthropology* 63(3): 350–59. https://doi.org/10.1086/720280.

BHP. 2018. *Water Report*. Retrieved 2 December 2023 from https://www.bhp.com/-/media/documents/investors/annual-reports/2018/180828_bhpwaterreport2018.pdf.

———. 2019. *Water Stewardship Position Statement*. Last modified 21 April 2022. Retrieved 2 December 2023 from https://www.bhp.com/sustainability/environment/water.

Billo, Emily. 2020. 'Gendering Indigenous Subjects: An Institutional Ethnography of Corporate Social Responsibility in Ecuador', *Gender, Place & Culture* 27(8): 1134–154. https://doi.org/10.1080/0966369X.2019.1650723.

Birkenholtz, Trevor L. 2015. 'Recentralizing Groundwater Governmentality: Rendering Groundwater and its Users Visible and Governable', *WIREs Water* 2 (1): 21–30. https://doi.org/10.1002/wat2.1058.

Blaser, Mario, Harvey A. Feit, and Glenn McRae (eds). 2004. *In the Way of Development: Indigenous Peoples, Life Projects and Globalization*. London: Zed Books.

Boelens, Rutgerd, and Paul H. Gelles. 2005. 'Cultural Politics, Communal Resistance and Identity in Andean Irrigation

Development', *Bulletin of Latin American Research* 24(3): 311–27. https://doi.org/10.1111/j.0261-3050.2005.00137.x.
Bolados García, Paola, and Alejandra Sánchez Cuevas. 2017. 'Una Ecología Política Feminista en Construcción: El Caso de las 'Mujeres de Zonas de Sacrificio en Resistencia', Región de Valparaíso, Chile', *Psicoperspectivas* 16: 33–42. Retrieved 17 September 2024 from http://www.scielo.cl/scielo.php?script=sci_arttext&pid=S0718-69242017000200033&nrm=iso.
Bolados, Paola. 2014. 'Los Conflictos Etnoambientales de 'Pampa Colorada' y 'El Tatio' en el Salar de Atacama, norte de Chile. Procesos Etnicos en un Contexto Minero y Turístico Transnacional', *Estudios Atacameños* 48: 229–48. http://dx.doi.org/10.4067/S0718-10432014000200015.
Bonelli, Cristóbal, and Cristina Dorador. 2021. 'Endangered Salares: Micro-disasters in Northern Chile', *Tapuya: Latin American Science, Technology and Society* 4(1). https://doi.org/10.1080/25729861.2021.1968634.
Borofsky, Robert. 2020. 'Rethinking Ethnography: A Study in Public Anthropology', *Anthropology Today* 36(5): 1–2. https://doi.org/10.1111/1467-8322.12600.
Briscoe, John, Pablo Anguita Salas, and Humberto Peña T. 1998. 'Managing Water as an Economic Resource: Reflections on the Chilean Experience', *World Bank Environment Department Papers* (62): 2–11.
Brown, Wendy. 2019. *In the Ruins of Neoliberalism: The Rise of Antidemocratic Politics in the West*. New York: Columbia University Press.
Budds, Jessica. 2012. 'Water, Power, and the Production of Neoliberalism in Chile, 1973–2005', *Environment and Planning D: Society and Space* 31: 301–18. https://doi.org/https://doi.org/10.1068/d9511.
Bustamante-Rivera, Gonzalo, and Thibault Matin. 2018. 'Beneficios Compartidos y la Gobernanza de la Extracción de Recursos Naturales en Territorios Indígenas: Aportes y Limitaciones para Latinoamérica', *Perfiles Latinoamericanos* 26(52). https://doi.org/10.18504/pl2652-009-2018.
Bustos, Beatriz, Mauricio Folchi, and Maria Fragkou. 2017. 'Coal Mining on Pastureland in Southern Chile; Challenging Recognition and Participation as Guarantees for Environmental Justice', *Geoforum* 84 (August): 292–304. https://doi.org/10.1016/j.geoforum.2015.12.012.
Bustos G., Beatriz, and Hernán Blanco P. no date. 'Patta Hoiri and Likanantay people: Rescuing the Knowledge of the Land'. *RIDES and Millenium Ecosystem Assessment Report* (Supplement). Retrieved 8 July 2015 from www.millenniumassessment.org/en/SGA.Chile.html.
Cabello, José. 2022. 'Reservas, Recursos y Exploración de Litio en Salares del Norte de Chile', *Andean Geology* 49(2): 297–306. http://dx.doi.org/10.5027/andgeov49n2-3444.
Cámera de Diputados, Gobierno de Chile. 2017. *Informe de la Comisión Especial Investigadora de los Actos de los Organismos Públicos Competentes, Encargados de la Fiscalización y Protección de Glaciares, Cuencas Hidrográficas y Salares de Chile*, 3 January. Retrieved 11 January 2017 from https://www.camara.cl/.

Cardenas, Luis. 2019. 'Minera Delfin: proyecto ligado a Hernan Buchi despliega lobby por permiso ambiental', *La Tercera*, 4 March 2019.
Carrasco, Anita. 2010. 'A Sacred Mountain and the Art of "Impression Management"', *Mountain Research and Development* 30(4): 391–97. https://doi.org/10.1659/MRD-JOURNAL-D-09-00065.1.
———. 2020. *Embracing the Anaconda: A Chronicle of Atacameño LIfe and Mining in the Andes*. Washington: Lexington Books.
Carrier, James G. 2018. 'Moral Economy: What's in a Name', *Anthropological Theory* 18(1): 18–35. http://doi/10.1177/1463499617735259.
Castro Rojas, Maria Victoria, and Varinia Varela Guarda. 1994. *Ceremonias de Tierra y Agua. Ritos Milenarios Andinos*. Santiago (Chile): Fondo de Desarrollo de la Cultura y las Artes.
Castro, Victoria, and Carlos Aldunate. 2003. 'Sacred Mountains in the Highlands of the South-Central Andes', *Mountain Research and Development* 23(1): 73–79. http://www.jstor.org/stable/3674538.
Castro-Lucic, Milka. 2002. 'Local Norms and Competition for Water in Aymara and Atacama Communities', in R. Boelens and P. Hoogendam (eds), *Water Rights and Empowerment*. Assen (The Netherlands): Van Goorcum, pp. 186–201.
Comisión de Evaluación (Antofagasta Region, Chile). 2019. *Resolucion de Desistimiento, 'Planta de Sulfato de Cobre Pentahidratado'*, Sistema de Evaluación de Impacto Ambiental. Retrieved 18 July 2019 from http://seia.sea.gob.cl/expediente/ficha/fichaPrincipal.php?modo=normal&id_expediente=7893580.
Comisión de Litio. 2015. *Informe Final. 'Litio: Una Fuente de energia, una oportunidad para Chile'*, Ministerio de Minería, Gobierno de Chile (Santiago).
Contardo, Esteban David. 2022. *Náusea: Crónica de una Zona de Sacrificio*. Santiago, Chile: La Pollera Ediciones.
Cubillos, Carolina F., et al. 2018. 'Microbial Communities From the World's Largest Lithium Reserve, Salar de Atacama, Chile: Life at High LiCl Concentrations', *Journal of Geophysical Research. Biogeosciences* 123 (12): 3668–681. https://doi.org/10.1029/2018JG004621.
Daston, Lorraine. 1995. 'The Moral Economy of Science', *Osiris* 10 (1): 2–24. https://doi.org/10.1086/368740.
de la Cadena, Marisol. 2010. 'Indigenous Cosmopolitics in the Andes: Conceptual Reflections beyond "Politics"', *Cultural Anthropology* 25(2): 334–70. https://doi.org/10.1111/j.1548-1360.2010.01061.x.
———. 2015. *Earth Beings: Ecologies of Practice across Andean Worlds*. Durham and London: Duke University Press.
Descola, Philippe. 2013. *Beyond Nature and Culture*, trans. J. Lloyd. Chicago and London: The University of Chicago Press.
Dingman, Robert J. 1967. 'Geography and Groundwater Resources of the Northern Part of the Salar de Atacama, Antofagasta Province, Chile', *Geological Survey Bulletin* 1219: 1–50.
Dirección General de Aguas (DGA). 1999. *Evaluación de la Disponibilidad de Recursos Hidricos para Constituir Derechos de Aprovechamiento en las*

Subcuencas Afluentes al Salar de Atacama. II Region. Direccion General de Aguas, Departamento de Administracion de Recursos Hídricos, Minuta Tecnica No. 60 (Santiago, May 1999). Retrieved 4 May 2017 from https://snia.mop.gob.cl/repositoriodga/handle/20.500.13000/4921.

———. 2010 *Actualización de la Evaluación de la Disponibilidad de Recursos Hídricos para Constituir Derechos de Aprovechamiento en las Subcuencas Afluentes al Salar de Atacama. II Region.* January, S.I.T No210 GCF Ingenieros Ltda. Santiago: Ministerio de Obras Publicas, Gobierno de Chile (Santiago). Retrieved 21 July 2016 from https://bibliotecadigital.ciren.cl/items/ee86aa8c-e170-4944-9485-1398e7e4b849.

———. 2012. 'Taller de presentación de resultados. Diagnostico plan Estratégico para la gestión de los recursos hídricos', public document. Ministerio de Obras Públicas, Gobierno de Chile. San Pedro de Atacama, 21 December.

———. 2014. *Diagnostico y Sistematizacion de Informacion de Planes de Alerta Temprana Vigentes con Condicionamiento de Derechos*. Informe Final Volumen No.1, April, Hidromas CEF Ltda., S.I.T 336, División de Estudios y Planificación. Santiago: Ministerio de Obras Públicas, Gobierno de Chile. Retrieved 4 May 2017 from https://repositoriodirplan.mop.gob.cl/biblioteca/handle/20.500.12140/25862.

Dolan, Catherine, and Dinah Rajak. 2016. 'Introduction: Toward the Anthropology of Corporate Social Responsibility', in Catherine Dolan and Dinah Rajak (eds), *The Anthropology of Corporate Social Responsibility*. London and New York: Berghahn, pp. 1–28.

Donoso, Guillermo, Elisabeth Lictevout, and Jean-Daniel Rinaudo. 2020. 'Groundwater Management Lessons from Chile', in Jean-Daniel Rinaudo, et al. (eds), *Sustainable Groundwater Management: A Comparative Analysis of French and Australian Policies and Implications to Other Countries*. Cham: Springer International Publishing, pp. 481–509.

Duranti, Alessandro. 2010. 'Husserl, Intersubjectivity and Anthropology', *Anthropological Theory* 10 (1–2): 16–35. https://doi.org/10.1177/1463499610370517.

Escobar, Arturo. 1995. *Encountering Development: The Making and Unmaking of the Third World*. Princeton, NJ: Princeton University Press.

———. 2012. *Territories of Difference: Place, Movements, Life, Redes*. Durham and London: Duke University Press.

———. 2017. 'Sustaining the Pluriverse: The Political Ontology of Territorial Struggles in Latin America', in Marc Brightman and Jerome Lewis (eds), *The Anthropology of Sustainability: Beyond Development and Progress*. New York: Palgrave Macmillan US, pp. 237–56.

———. 2018. *Designs for the Pluriverse: Radical Interdependence, Autonomy and the Making of Worlds*. Durham: Duke University Press.

Falleti, Tulia G., and Thea N. Riofrancos. 2018. 'Endogenous Participation: Strengthening Prior Consultation in Extractive Economies', *World Politics* 70(1): 86–121. https://doi.org/10.1017/S004388711700020X.

Fassin, Didier. 2011. 'A Contribution to the Critique of Moral Reason', *Anthropological Theory* 11(4): 481–91. https://doi.org/10.1177/1463499611429901.
Ferguson, James. 1994. *The Anti-Politics Machine: 'Development,' Depoliticization, and Bureaucratic Power in Lesotho*. Minneapolis: University of Minnesota Press.
———. 2010. 'The Uses of Neoliberalism', *Antipode* 41: 166–84. https://doi.org/10.1111/j.1467-8330.2009.00721.x.
Finn, Janet L. 1998. *Tracing the Veins: Of Copper, Culture and Community from Butte to Chuquicamata*. Berkeley: University of California Press.
Folchi Donoso, Mauricio. 2003. 'La Insustentabilidad de *Boom* Minero Chileno: Politica y Medio Ambiente, 1983–2003', *Ecologia Politica* 26: 23-–49. https://doi.org/https://repositorio.uchile.cl/handle/2250/122792.
Fortun, Kim. 2012. 'Ethnography in Late Industrialism', *Cultural Anthropology* 27(3): 446–64. https://doi.org/10.1111/j.1548-1360.2012.01153.x.
———. 2014. 'From Latour to Late Industrialism', *Hau: Journal of Ethnographic Theory* 4(1): 309–29. http://dx.doi.org/10.14318/hau4.1.017.
Frazier, Lessie Joe. 2007. *Salt in the Sand: Memory, Violence and the Nation-State in Chile, 1890 to the Present*. Durham: Duke University Press.
Frederiksen, Tomas, and Matthew Himley. 2020. 'Tactics of Disposession: Access, Power, and Subjectivity at the Extractive Frontier', *Transactions of the Institute of the British Geographical Society* 45: 50–64. https://doi.org/10.1111/tran.12329.
Gajardo, Anahy. 2021. 'Performing the "India Permitida": The Counter-Gift of Indigenous Women Targeted by a Corporate Social Responsibility Program (Chile)', Bulletin of Latin American Research 40(2): 172–87. https://doi.org/10.1111/blar.13143.
Gajardo, Gonzalo, and Stella Redón. 2019. 'Andean Hypersaline Lakes in the Atacama Desert, northern Chile: Between Lithium Exploitation and Unique Biodiversity Conservation', *Conservation Science and Practice* 1(9). https://doi.org/10.1111/csp2.94.
Galeano, Eduardo. 1971. *Las Venas Abiertas de America Latina*. Mexico: Siglo XXI Editores.
Gamu, Jonathan Kishen, and Peter Dauvergne. 2018. 'The Slow Violence of Corporate Social Responsibility: The Case of Mining in Peru', *Third World Quarterly* 39(5): 959–75. https://doi.org/10.1080/01436597.2018.1432349.
Garcia, Magdalena B. 2021. *Otra Costa Hay en la Puna: Configuraciones del Espacio Indígena y Movilidad Pastoril en los Andes Centro-Sur*. Santiago: Qillqa Ediciones.
Garcia, Magdalena B., Manuel Prieto, and Fernanda Kalazich. 2021. 'The Protection of the Mountain Ecosystems of the Southern Central Andes: Tensions Between Aymara Herding Practices and Conservation Policies', *Journal on Protected Mountain Areas Research and Management* 13(1): 22–30. https://doi.org/10.1553/eco.mont-13-1s22.
Gardner, Katy, and David Lewis. 2015. *Anthropology and Development: Challenges for the Twenty-First Century*, 2nd edn, vol. 16. London: Pluto Press.
Garsten, Christina, and Kerstin Jacobsson. 2011. 'Transparency and Legibility in International Institutions: The UN Global Compact and Post-Political Global

Ethics', *Social Anthropology* 19(4): 378–93. https://doi.org/10.1111/j.1469-8676.2011.00171.x.
Gedda, Francisco. (dir). 1985. *El Altiplano: Las Alturas Vivies de los Andes* (Film recording, held in Biblioteca Museo Chileno de Arte Precolombino, Santiago, Chile).
———. (dir). 2010. Atacameños, Vida Social y Costumbres, *Colección Pueblos Originarios*. Chile: Sur Imagen.
Gelles, Paul H. 2002. *Agua y Poder en la Sierra Peruana*. Lima: Pontifica Universidad Catolica del Peru.
Gershon, Ilana. 2011. 'Neoliberal Agency', *Current Anthropology*. 52(4): 537–55. R https://doi.org/10.1086/660866.
Giddens, Anthony. 1991. *Self and Society in a Late Modern Age*. Stanford: Stanford University Press.
Gilberthorpe, Emma, and Dinah Rajak. 2016. 'The Anthropology of Extraction: Critical Perspectives on the Resource Curse', *The Journal of Development Studies*: 1–19. https://doi.org/10.1080/00220388.2016.1160064.
Gleeson, Tom, et al. 2020. 'Global Groundwater Sustainability, Resources, and Systems in the Anthropocene', *Annual Review of Earth and Planetary Sciences* 48(1): 431–63. https://doi.org/10.1146/annurev-earth-071719-055251.
Göbel, Barbara. 2001. 'Risk and Culture in the Andes: Differences Between Indigenous and Western Development Perspectives', in Gisela Böhm (ed.), *Environmental Risks: Perception, Evaluation and Management*. Oxford: Oxford University Press.
———. 2008. 'Dangers, Experience and Luck: Living with Uncertainty in the Andes', in Michael J. Casimir (ed.), *Culture and the Changing Environment: Uncertainty, Cognition and Risk Management in Cross Cultural Perspective*. New York: Berghahn Books, pp. 221–50.
Gobierno [Militar] de Chile. c.1990. *Memoria de Gobierno 1973–1990. Tomo II Economico Productivo*. Santiago, Chile.
Gobierno de Chile. 2022a. *Estrategia Nacional del Litio: Por Chile y su Gente*. Santiago, Chile: CORFO.
———. 2022b. *Proyecto de Ley Que Reforma El Código de Aguas* [Water Code Amendment Bill] (Chile), boletín N° 7.543–12.
Golub, Alex. 2014. *Leviathans at the Gold Mine: Creating Indigenous and Corporate Actors in Papua New Guinea*. Durham and London: Duke University Press.
Gómez-Barris, Macarena. 2017. *The Extractive Zone: Social Ecologies and Decolonial Perspectives*. Durham: Duke University Press.
Gonzáles, Cecilio, et al. 2014. *Así Pasa Cuando Sucede: Testimonios de Vida Atacameños*. Paris: Editions du Relief.
Goodale, Mark. 2016. 'Dark Matter: Toward a Political Economy of Indigenous Rights and Aspirational Politics', *Critique of Anthropology* 36(4): 439–57. https://doi.org/10.1177/0308275X15619017.
Goodale, Mark, and Nancy Postero. 2013. *Neoliberalism, Interrupted: Social Change and Contested Governance in Latin America*. Stanford: Stanford University Press.

Grebe, María Ester. 2004. 'Amerindian Music of Chile', in Malena Kuss (ed.), *Music in Latin America and the Caribbean: An Encyclopedic History. Volume 1 'Performing Beliefs', Indigenous Peoples of South America, Central America and Mexico*. Austin: University of Texas Press.

Greenberg, Jessica, and Sarah Muir. 2022. 'Disappointment', *Annual Review of Anthropology* 51(1): 307–23. https://doi.org/10.1146/annurev-anthro-041520-105355.

Gregory, Chris and Jon Altman (eds). 2018. *The Quest for the Good Life in Precarious Times: Ethnographic Perspectives on the Moral Economy*. ANU E-Press. http://doi.org/10.22459/QGLPT.03.2018.

Gudynas, Eduardo. 2009. 'Diez Tesis Urgentes sobre el Nuevo Extractivismo: Contextos y Demandas Bajo el Progresismo Sudamericano Actual', in Jürgen Schuldt, et al. (eds), Extractivismo, Política y Sociedad. Quito: Centro Andino de Acción Popular, pp. 187–225.

———. 2012. 'Estado Compensador y Nuevos Extractivismos. Las Ambivalencias del Progresismo Sudamericano', *Nueva Sociedad* 237: 128–46. https://doi.org/https://nuso.org/articulo/estado-compensador-y-nuevos-extractivismos-las-ambivalencias-del-progresismo-sudamericano/.

Gundermann, Hans, and Barbara Göbel. 2018. 'Comunidades Indígenas, Empresas del Litio y Sus Relaciones en el Salar de Atacama', *Chungara Revista de Antropología Chilena* 50 (3): 471–86. http://dx.doi.org/10.4067/S0717-73562018005001602.

Gundermann Kröll, Hans. 2002. *Los Atacameños del Siglo XIX y XX, Una Antropología Historica Regional. Documento de Divulgación Solicitado por el Subgrupo de Trabajo Pueblo Atacameño, Grupo de Trabajo Pueblos Indígenas del Norte*. San Pedro de Atacmana: Comisión de Verdad Histórica y Nuevo Trato.

Guzmán, Patricio. (dir.) 2010. *Nostalgia de la Luz* [Film]. Chile and Spain: Pyramide International, Icarus Films.

Haraway, Donna. 2016. *Staying with the Trouble: Makin kin in the Chthulucene*. Durham: Duke University Press.

Harris, Olivia. 1989. 'The Earth and the State: The Sources and Meanings of Money in Northern Potosi', in Maurice Bloch and Jonathon Parry (eds), *Money and the Morality of Exchange*. Cambridge: Cambridge University Press, pp. 232–68.

Harvey, David. 2005. *Brief History of Neoliberalism*. Oxford: Oxford University Press.

Harvey, Penny, and Hannah Knox. 2014. 'The Enchantments of Infrastructure', in D. Dalakoglou and P. Harvey. *Roads and Anthropology: Ethnography, Infrastructures, (Im)mobility*, London: Routledge [online].

Hegel, Georg Wilhelm Fredrich. 1953. *Reason in History. A General Introduction to the Philosophy of History*. New York: MacMillan.

Hidalgo B., and Blas E. 1992. 'Organización Social, Tradición y Aculturación en Socaire, una aldea Atacameña', dissertation. Santiago: Universidad de Chile, Departamento de Antropología.

Houston, John. 2006. 'Evaporation in the Atacama Desert: An Empirical Study of Spatio-Temporal Variations and Their Causes', *Journal of Hydrology* 330 (3–4): 402–12. http://dx.doi.org/10.1016/j.jhydrol.2006.03.036.

Jenkins, Heledd. 2004. 'Corporate Social Responsibility and the Mining Industry: Conflicts and Constructs', *Corporate Social Responsibility and Environmental Management* 11(1): 23–34. https://doi.org/10.1002/csr.50.

Jerez, Bárbara, Ingrid Garcés, and Robinson Torres. 2021. 'Lithium Extractivism and Water Injustices in the Salar de Atacama, Chile: The Colonial Shadow of Green Electromobility', *Political Geography* 87: 102382. https://doi.org/10.1016/j.polgeo.2021.102382.

Jerez Henriquez, Bárbara. 2018. *Impacto Socioambiental de la Extracción de Litio en las Cuencas de los Salares Altoandinos del Cono Sur*. Observatorio de Conflictos Mineros de América Latina (Santiago).

Kalazich, Fernanda. 2015. 'Memory as Archaeology. An Experience of Public Archaeology in the Atacama Desert', *Public Archaeology* 14 (1): 44–65. https://doi.org/10.1179/1465518715Z.00000000086.

———. 2013. 'Cultural Meanings and Values of the Past. A Participatory Approach to Archaeology in the Atacameño Community of Peine, Chile', Ph.D. dissertation. London: Institute of Archaeology, University College London.

Kampf, Stephanie K., et al. 2005. 'Evaporation and Land Surface Energy Budget at the Salar de Atacama, Northern Chile', *Journal of Hydrology* 310 (1–4): 236–52. https://doi.org/http://dx.doi.org/10.1016/j.jhydrol.2005.01.005.

Karliner, Joshua. 1997. *The Corporate Planet: Ecology and Politics in the Age of Globalization*. San Francisco: Sierra Club Books.

Katwala, A. 2019. 'The Devastating Environmental Impact of Technological Progress', *Wired*, 27 October, no. Environment. Retrieved 20 July 2020 from https://doi.org/wired.co.uk/article/lithium-copper-mining-atacama-desert.

Kaunda, Rennie B. 2020. 'Potential Environmental Impacts of Lithium Mining', *Journal of Energy & Natural Resources Law* 38(3): 237–44. https://doi.org/10.1080/02646811.2020.1754596.

Kelty, Christopher M. 2017. 'Too Much Democracy in All the Wrong Places. Toward a Grammar of Participation', *Current Anthropology* 58 (S15): S77–S90. https://www.jstor.org/stable/26544378.

Kirsch, Stuart. 2010. 'Sustainable Mining', *Dialectical Anthropology* 34(1): 87–93. https://doi.org/10.1007/s10624-009-9113-x.

———. 2014. *Mining Capitalism: The Relationship Between Corporations and their Critics*. Berkeley: University of California Press.

Korf, Benedikt. 2023. 'The Irony of Development: Critique, Complicity, Cynicism', *Anthropological Theory* 23(2): 147–66. https://doi.org/10.1177/14634996221115225.

Langton, Marcia, and Odette Mazel. 2008. 'Poverty in the Midst of Plenty: Aboriginal people, the Resource Curse and Australia's Mining Boom', *Journal of Energy and Natural Resources Law* 26(1): 31–65. https://doi.org/10.1080/02646811.2008.11435177.

Lambek, Michael. 2011. 'Catching the Local', *Anthropological Theory* 11(2): 197–221. https://doi.org/10.1177/1463499611407397.
Latcham, Ricardo A. 1926. *Chuquicamata, estado yankee (vision de la montaña roja).* Santiago, Chile: Editorial Nascimiento.
Latta, Alex, and Beatriz E. Cid Aguayo. 2012. 'Testing the Limits: Neoliberal Ecologies from Pinochet to Bachelet', *Latin American Perspectives* 39(4): 163–80.
Lemus A., Ana María. 2019. 'Cuestión de Altura. Turismo en Diferentes Escalas en el Oasis de San Pedro de Atacama', in Francisca De la Maza (ed.), *Turismo en Territorios Indígenas: Identidad, Resistencia y Poder*. Santiago: CIIR and Pehuén, pp. 51–81.
Li, Fabiana. 2011. 'Engineering Responsibility: Environmental Mitigation and the Limits of Commensuration in a Chilean Mining Project', *Focaal* 2011 (60): 61–73. https://doi.org/10.3167/fcl.2011.600106.
———. 2015. *Unearthing Conflict: Corporate Mining, Activism and Expertise in Peru.* Durham (NC): Duke University Press.
Li, Tanya Murray. 2014. *Land's End: Capitalist Relations on an Indigenous Frontier.* Durham: Duke University Press.
Linton, Jamie, and Jessica Budds. 2014. 'The Hydrosocial Cycle: Defining and Mobilizing a Relational-Dialectical Approach to Water', *Geoforum* 57: 170–80. http://dx.doi.org/10.1016/j.geoforum.2013.10.008.
Liu, Wenjuan, and Datu B. Agusdinata. 2020. 'Interdependencies of Lithium Mining and Communities Sustainability in Salar de Atacama, Chile', *Journal of Cleaner Production* 260: 120–38. https://doi.org/https://doi.org/10.1016/j.jclepro.2020.120838.
Livingstone, G. 2019. 'The Farmers Who Worry about Our Phone Batteries', *BBC News*, 15 August 2019, no. Business. Retrieved 1 September 2019 from https://doi.org/bbc.com/news/business-49355817.
Luisetti, Federico. 2019. 'Geopower: On the States of Nature of Late Capitalism', *European Journal of Social Theory* 22(3): 342–63. https://doi.org/10.1177/1368431018803764.
Lyons, Kristen, and Carol Richards. 2013. 'Mining Universities: Can We Count on the Integrity of Research Coming out of Today's Mining-Funded Tertiary Sector?' *Arena Magazine* 124: 7–9.
Macpherson, Elizabeth. 2017. 'Beyond Recognition: Lessons from Chile for Allocating Indigenous Water Rights in Australia', *UNSW Law Journal* 40(3): 1130–169.
———. 2019. *Indigenous Water Rights in Law and Regulation: Lessons from Comparative Experience*. Cambridge: Cambridge University Press.
Macpherson, Elizabeth, et al. 2023. 'Evolving Rights to (and of) Water in Chile: a Case for Relationship-Based Water Law and Governance', *The International Journal of Human Rights*: 1–26. https://doi.org/10.1080/13642987.2023.2266719.
Mahlman, Anst von H. 1854. *Desierto de Atacama* [Map]. Berlin: Geo-lith Anst v. H. Mahlman.
Marazuela, Miguel A., et al. 2020. 'Towards More Sustainable Brine Extraction in Salt Flats: Learning from the Salar de Atacama',

Science of The Total Environment 703: 135605. https://doi.org/https://doi.org/10.1016/j.scitotenv.2019.135605.
Marazuela, Miguel A., et al. 2021. 'La Importancia de Incorporar la Hidrodinámica de la Interfaz Salina en la Gestión de los Recursos Minerales y Ecosistemas de los Salares', *Boletín Geológico y Minero* 132 (1–2): 127–39. http://doi.org/10.21701/bolgeomin.132.1-2.013.
Marzec, Robert P. 2018. 'Securing the Future in the Anthropocene: A Critical Analysis of the Millennium Ecosystem Assessment Scenarios', *Elementa: Science of the Anthropocene* 6. https://doi.org/10.1525/elementa.294.
Maturana, Humberto R., and Francisco J. Varela. 1980. *Autopoiesis and Cognition*. Dordrecht: Springer.
McGoey, Lindsey. 2012. 'Strategic Unknowns: Towards a Sociology of Ignorance', *Economy and Society* 41(1): 1–16.
McQuarrie, Michael. 2013. 'No Contest: Participatory Technologies and the Transformation of Urban Equality', *Public Culture* 25(1): 143–75. https://doi.org/10.1215/08992363-1890495.
Mejia-Muñoz, Sara, and Sally Babidge. 2023. 'Lithium Extractivism: Perpetuating Historical Asymmetries in the 'Green economy', *Third World Quarterly* 44(6): 1119–136. https://doi.org/10.1080/01436597.2023.2176298.
Mercado Guerra, Javier. 2007. 'Los Inicios de la chilenización en Atacama: Una Aproximación a las Discursivedades sobre el "Indio Atacameño" durante la Posguerra del Pacífico (1885–1910)', ResearchGate. Retrieved 17 September 2024 from https://www.researchgate.net/publication/333390496.
Merino, Roger. 2018. 'Re-Politicizing Participation or Reframing Environmental Governance? Beyond Indigenous' Prior Consultation and Citizen Participation', *World Development* 111: 75–83. https://doi.org/https://doi.org/10.1016/j.worlddev.2018.06.025.
Messerli, Bruno, Martin Grosjean, and Mathias Vuille. 1997. 'Water Availability, Protected Areas, and Natural Resources in the Andean Desert Altiplano', *Mountain Research and Development* 17(3): 229–38. https://www.jstor.org/stable/3673850.
Minera Escondida Limitada/ BHP. c.1997. Peine: 'Esfuerzo de Sus Habitantes' [Film]. (Collection of the Comunidad Indigena de Peine).
Minera Escondida Limitada. 2006. *Reporte de Sustentabilidad 2006 [Sustainability Report]*. Minera Escondida Limitada, operated by BHP Billiton.
———. 2010. *Plan de Alerta Temprana para el acuifero Monturaqui—Negrillar—Tilopozo. Revision de los datos de monitoreo, Informe No. 9. Revision hasta fines de 2009*. BHP Billiton, Schlumberger Water Services Chile Ltda (Antofagasta).
Ministerio de Justicia Republica de Chile. 2008. *Código de Aguas [con las modificaciones]*. Santiago: Biblioteca Nacionale de Chile. Retrieved 9 June 2015 from http://www.leychile.cl/Navegar/?idNorma=5605&idVersion=2008-12-13&idParte#indigena0.
Molina, Raúl Otárola. 2017. *'En la Huella está el Peligro': Arrierías de Puna y Desierto*. San Pedro de Atacama (Chile): Qillqa Ediciones.

Molle, François, and Alvar Closas. 2019. 'Groundwater Governance', in P. Maurice (ed.), *Encyclopedia of Water: Science, Technology and Society*. https://doi.org/10.1002/9781119300762.wsts0191.

Mora Rodriguez, Oriana and Juan Cayo Rivera. 2017. 'Una Breve Historia del Agua, el Bien Natural de los Lickanantay, Defensa y Amenazas en un Contexto de Administración-Protección en el Salar de Atacama, en el norte de Chile (Ponencia)', Congreso El Extractivismo en América Latina: Dimensiones Económicas, Sociales, Políticas y Culturales 2017: 58–75.

Moran, Theodore H. 1974. *Multinational Corporations and the Politics of Dependence: Copper in Chile*. Princeton: Princeton University Press.

Moran, Brendan J., et al. 2022. 'Relic Groundwater and Prolonged Drought Confound Interpretations of Water Sustainability and Lithium Extraction in Arid Lands', *Earth's Future* 10(7): e2021EF002555. https://doi.org/https://doi.org/10.1029/2021EF002555.

Morgan, Richard K. 2012. 'Environmental Impact Assessment: the State of the Art', *Impact Assessment and Project Appraisal* 30 (1): 5–14. https://doi.org/10.1080/14615517.2012.661557.

Mostny, Grete. 1954. *Peine: Un Pueblo Atacameño*. Facultad de Filosofia. Universidad de Chile, Santiago: Publicacion No.4 Del Instituto de Geografia.

Moulian, Tomás. 1998. *Chile Actual: Anatomía de un Mito*. Santiago (Chile): LOM Ediciones, ARCIS Universidad.

Nixon, Rob. 2011. *Slow Violence and the Environmentalism of the Poor*. Cambridge, MA: Harvard University Press.

North, Liisa L., and Ricardo Grinspun. 2016. 'Neo-Extractivism and the New Latin American Developmentalism: The Missing Piece of Rural Transformation', *Third World Quarterly* 37(8): 483–1504. https://doi.org/10.1080/01436597.2016.1159508.

Nuñez Henriquez, Patricio. 2006 [1896]. *Glossorio de la Lengua Atacameña* (Emilio F. Vaïsse-Felix and Anibal Echeverria Y Reyes [eds]). Antofagasta: Universidad De Antofagasta.

Núñez Srýtr, Marinka. 1998a. *'Eso yo vi, en esos años de la gente Antigua. . .' Memorias y olvidos'*. Santiago: MN. Practical Professional Antropología Universidad Academia. Humanismo Cristiano.

———. 1998b. 'Peine: Saber andino, manejo de recursos y transformaciones', *Estudios Atacameños* (16): 283–92. http://www.jstor.org/stable/25674722.

———. 2011. 'Rutas, viajes y convidos: territorialidad peineña en las cuencas de Atacama y Punta Negra', in Lautaro Núñez A. and Axel E. Nielsen (eds), *En Ruta: Arqueología, Historia y Etnografía del Trafico Sur Andino*. Córdoba: Encuentro Grupo Editor, pp. 373–98.

———. 2015. 'Sociedad, Naturaleza y Territorialidad en el Desierto y Puna de Atacama. Siglos XX–XXI', Ph.D. dissertation. San Pedro de Atacama (Chile): Instituto de Investigaciones Arqueologicas y Departamento de Antropología Universidad Catolica del Norte y Universidad de Tarapacá.

O'Faircheallaigh, Ciaran, and Sally Babidge. 2023. 'Negotiated Agreements, Indigenous Peoples and Extractive Industry in the Salar de Atacama, Chile:

When Is an Agreement More than a Contract?' *Development and Change* 54(3): 641–70. https://doi.org/https://doi.org/10.1111/dech.12767.

O'Faircheallaigh, Ciaran. 2016. *Negotiations in the Indigenous World: Aboriginal Peoples and Extractive Industry in Australia and Canada*. New York: Routledge.

———. 2023. *Indigenous Peoples and Mining: A Global Perspective*. Oxford: Oxford University Press.

Odell, Scott D. 2021. 'Hydrosocial Displacements: Sources and Impacts of Collaboration as a Response to Water Conflict near three Chilean Mines', *Resources Policy* 74: 102305. https://doi.org/10.1016/j.resourpol.2021.102305.

Ojeda, José M. Figueroa. 2002. 'Minería y Responsibilidad Social Corporativa: El Caso de Fundacion Minera Escondida', in *VI Encuentro de seguridad y medio ambiente*. [Conference Proceedings].

Orlove, Ben, and Steven C. Caton. 2010. 'Water Sustainability: Anthropological Approaches and Perspectives', *Annual Review of Anthropology* 39: 401–15.

Ostry, Jonathan D., Prakash Loungani, and Davide Furceri. 2016. 'Neoliberalism: Oversold?' *Finance & Development* 53(2): 38–41. Retrieved 17 September 2024 from http://www.imf.org/external/pubs/ft/fandd/2016/06/pdf/ostry.pdf.

Owen, John R., and Deanna Kemp. 2012. 'Social Licence and Mining: a Critical Perspective', *Resources Policy* 16 (6). https://doi.org/https://doi.org/10.1016/j.resourpol.2012.06.016.

———. 2017. *Extractive Relations: Countervailing Power and the Global Mining Industry*. London: Routledge.

Paley, J. 2001. *Marketing Democracy: Power and Social Movements in Post-Dictatorship Chile*. Berkeley: University of California Press.

Parry, J, and M Bloch. 1989. 'Introduction: Money and the Morality of Exchange', in Jonathan Parry and Maurice Bloch (eds), *Money and the Morality of Exchange*. Cambridge: Cambridge University Press.

Philippi, Rodolfo. 1860. *Viaje al Desierto de Atacama Hecho de Orden del Gobierno de Chile en el Verano 1853–54*. Halle (Germany): Eduard Anton.

Povinelli, Elizabeth, A. 2002. *The Cunning of Recognition: Indigenous Alterities and the Making of Australian Multiculturalism*. Durham: Duke University Press.

———. 2011. *Economies of Abandonment: Social Belonging and Endurance in Late Liberalism*. Duke University Press.

———. 2016. *Geontologies: A Requiem to Late Liberalism*. Durham: Duke University Press.

Prieto, Manuel. 2015. 'Privatizing Water in the Chilean Andes: The Case of Las Vegas de Chiu Chiu', *Mountain Research and Development* 35(3): 220–29. https://doi.org/10.1659/MRD-JOURNAL-D-14-00033.1.

Prieto, Manuel, María Christina Fragkou, and Matías Calderón. 2020. 'Water Policy and Management in Chile', in Patricia Maurice (ed.), *Encyclopedia of Water: Science, Technology and Society*. West Sussex: Wiley, p. 11.

Prigogine, I., and Isabelle Stengers. 1984. *Order Out of Chaos: Man's New Dialogue with Nature*. Boulder, CO: New Science Library. Distributed by Random House.

Primer Tribunal Ambiental. 2020. *First Environmental Tribunal receives legal suit for environmental damage against Minera Escondida*. Primer Tribunal Ambiental. Retrieved 23 April 2020 from https://www.1ta.cl/.

Prno, Jason, and D. Scott Slocombe. 2012. 'Exploring the Origins of "Social License to Operate" in the Mining Sector: Perspectives from Governance and Sustainability Theories', *Resources Policy* 37: 346–57. https://doi.org/10.1016/j.resourpol.2012.04.002.

Programa de Reduccion de Riesgos y Desastres—Unidad de Redes Transdisciplinarias, Vicerectoria de Investigacion y Desarrollo, Universidad de Chile. 2020. *Position Paper, Los territorios que habita(re)mos: ¿Que futuro existe para las zonas de sacrificio?* Universidad de Chile (Santiago, Chile).

Rajak, Dinah. 2011a. *In Good Company: An Anatomy of Corporate Social Responsibility*. Stanford: Standford University Press.

———. 2011b. 'Theatres of Virtue: Collaboration, Consensus, and the Social Life of Corporate Social Responsibility', *Focaal* 2011(60): 9–20. https://doi.org/10.3167/fcl.2011.600102.

RIDES. 2005. *Millenium Ecosystem Assessment: Human Well-being and Sustainable Management in San Pedro de Atacama, Chile – Executive Summary*. Santiago, Chile. Retrieved 8 July 2015 from http://www.millenniumassessment.org/en/SGA.Chile.html.

Riofrancos, T. 2019. 'What Green Costs', *Logicmag*, no. Nature (December 7 2019). Retrieved 2 June 2021 from https://doi.org/logicmag.io/nature/what-green-costs/.

Rivera, Francisco Javier. 2004. 'Anti-Flexibilizacion con Identidad/Alteridad. La Cuestion Etnia Atacamena Contemporanea y su Contexto Neoliberal-Flexible', *Revista de Antropologia Experimental* 4: 1–17.

Rodríguez Ríos, Roberto, and Boris Fica Gallardo. 2020. *Guía de Campo: Plantas Vasculares Acuáticas en Chile*. Concepción: Corporación Chilena de la Madera (CORMA).

Roitman, Janet. 2013. *Anti-Crisis*. Durham: Duke University Press.

Ruiz Encina, Carlos, and Sebastián Caviedes Hamuy. 2020. 'Estructura y Conflicto Social en la Crisis del Neoliberalismo Avanzado Chileno', *Espacio Abierto* 29 (1): 86–101. Retrieved 17 September 2024 from https://repositorio.uchile.cl/handle/2250/175352.

Sahlins, M.D. 1993. 'Goodbye to Tristes Tropes: Ethnography in the Context of Modern World', *Journal of Modern History* 65(1): 1–25.

Salazar, Noel B. 2013. 'Imagining Mobility at the "End of the World"', *History and Anthropology* 24(2): 233–52. https://doi.org/10.1080/02757206.2013.761211.

Sanhueza, Maria Cecilia. 2001. 'Las Poblaciones de la Puna de Atacama y su Relacion con los Estados Nacionales: Una Lectura desde el Archivo', *Revista de Historia Indígena*.

Sassen, Saskia. 2014. *Expulsions: Brutality and Complexity in the Global Economy*. Cambridge, MA: The Belknap Press of Harvard University Press.

Sawyer, Suzana. 2004. *Crude Chronicles: Indigenous Politics, Multinational Oil, and Neoliberalism in Ecuador*. Durham: Duke University Press.

Scott, James C. 1998. *Seeing Like a State: How Certain Schemes to Improve the Human Condition Have Failed.* New Haven, CT: Yale University Press.

Senate Standing Committee on Foreign Affairs Defence and Trade. 1992. *Australia and Latin America: Report of the Senate Standing Committee on Foreign Affairs Defence and Trade.* Commonwealth of Australia (Canberra).

Sherwood, Dave. 2020. 'Top Lithium Miner Seeks to Monitor Water Scarcity in Parched Chile Salt Flat', *Reuters (online)*, 10 February 2020, Sustainable Business.

Shever, Elana. 2012. *Resources for Reform: Oil and Neoliberalism in Argentina.* Stanford (CA): Stanford University Press.

Silva, Eduardo. 2008. *Challenging Neoliberalism in Latin America.* Cambridge: Cambridge University Press.

Smith, Gavin. 1989. *Livelihood and Resistance: Peasants and the Politics of Land in Peru.* Berkeley: University of California Press.

Smith, Jessica, and Frederico Helfgott. 2010. 'Flexibility or Exploitation: Corporate Social Responsibility and the Perils of Universalism', *Anthropology Today* 26(3): 20–23.

SQM. 2019. *Sustainability Report.* Retrieved 3 February 2021 from www.sqm.com/wp-content/uploads/2019/11/SQM-Sustainable-Lithium_ESPAÑOL-1.pdf.

Starn, Orin. 1994. 'Rethinking the Politics of Anthropology: The Case of the Andes', *Current Anthropology* 35(1): 13–38.

Stern, Steve J. 2006. *Battling for Hearts and Minds: Memory Struggles in Pinochet's Chile, 1973–1988.* Durham: Duke University Press.

Stevenson, Lisa. 2014. *Life Beside Itself. Imagining Care in the Canadian Arctic.* Berkeley: University of California Press.

Strathern, Marilyn. 2020. *Relations: An Anthropological Account.* Durham: Duke University Press.

Svampa, Maristella. 2019. *Neo-extractivism in Latin America: Socio-Environmental Conflicts, the Territorial Turn, and New Political Narratives.* Cambridge: Cambridge University Press.

Swyngedouw, Erik. 2009. 'The Political Economy and Political Ecology of the Hydro-Social Cycle', *Journal of Contemporary Water Research & Education* 142(1): 56–60. https://doi.org/10.1111/j.1936-704X.2009.00054.x.

Taussig, Michael T. 1997. *The Magic of the State.* New York: Routledge.

Tecklin, David, Carl Bauer, and Manuel Prieto. 2011. 'Making Environmental Law for the Market: the Emergence, Character, and Implications of Chile's Environmental Regime', *Environmental Politics* 20 (6): 879–98. https://doi.org/10.1080/09644016.2011.617172.

Tedlock, Barbara. 1991. 'From Participant Observation to the Observation of Participation: The Emergence of Narrative Ethnography', *Journal of Anthropological Research* 47(1): 69–94. www.jstor.org/stable/3630581.

Tejeda, Ignacio, Rodrigo Cienfuegos, Jose F. Muñoz, and Mario Durán. 2003. 'Numerical Modeling of Saline Intrusion in Salar de Atacama',

Journal of Hydrological Engineering 8 (1): 25–34. https://doi.org/https://doi.org/10.1061/(ASCE)1084-0699(2003)8:1(25).

Thiel, Vera, et al. 2010. 'Unique Communities of Anoxygenic Phototrophic Bacteria in Saline Lakes of Salar de Atacama (Chile): Evidence for a New Phylogenetic Lineage of Phototrophic Gammaproteobacteria from pufLM Gene Analyses', *FEMS Microbiology Ecology* 74(3): 510–22. https://doi.org/10.1111/j.1574-6941.2010.00966.x.

Trawick, Paul. 2001. 'The Moral Economy of Water: Equity and Antiquity in the Andean Commons', *American Anthropologist* 103 (2): 361–79. https://doi.org/https://doi.org/10.1525/aa.2001.103.2.361.

Trouillot, Michel-Rolph. 2001. 'The Anthropology of the State in the Age of Globalization', *Current Anthropology* 42(1): 125–38. https://doi.org/10.1086/318437.

Tsing, Anna Lowenhaupt. 1993. *In the Realm of the Diamond Queen: Marginality in an Out-of-the-way Place*. Princeton: Princeton University Press.

———. 2005. *Friction: An Ethnography of Global Connection*. Princeton: Princeton University Press.

TVN, 24 Horas. 2007. *Pampa Colorada*. Retrieved 20 May 2014 from https://www.youtube.com/ watch?v=DukwOsnUtHY.

United Nations. 2022. *United Nations World Water Development Report 2022: Groundwater: Making the Invisible Visible*. UNESCO. Retrieved 1 September 2023 from unesdoc.unesco.org/ark:/48223/pf0000380721 (Paris).

Ureta, Sebastián. 2018. 'Baselining Pollution: Producing 'Natural Soil' for an Environmental Risk Assessment Exercise in Chile', *Journal of Environmental Policy & Planning* 20(3): 342–55. https://doi.org/10.1080/1523908X.2017.1410430.

Valdivia Ortiz de Zárate, Verónica 2010. ''¡Estamos en Guerra Senores!'. El Regimen Militar de Pinochet y el 'Pueblo' 1973–1980', *Historia* 1(43): 163–201. Retrieved 17 September 2024 from https://www.redalyc.org/pdf/334/33416405005.pdf.

Valenzuela, América, and Ricardo Moyano. 2021. 'Ethnicity and Ritual in the Atacameños Andes: Water, Mountains, and Irrigation Channels in Socaire (Atacama, Chile)', in John E. Staller (ed.), *Andean Foodways: Pre-Columbian, Colonial, and Contemporary Food and Culture*. Cham: Springer International Publishing, pp. 423–45.

Vásquez, Carlos, et al. 2013. 'Modeling Flow and Reactive Transport to Explain Mineral Zoning in the Atacama Salt Flat Aquifer, Chile', *Journal of Hydrology* 490: 114–25. http://dx.doi.org/10.1016/j.jhydrol.2013.03.028.

Velásquez, Teresa A. 2012. 'The Science of Corporate Social Responsibility (CSR): Contamination and Conflict in a Mining Project in the Southern Ecuadorian Andes', *Resources Policy* 37(2): 233–40. http://dx.doi.org/10.1016/j.resourpol.2011.10.002.

Vilches, Flora. 2011. 'From Nitrate Town to Internment Camp: the Cultural Biography of Chacabuco, northern Chile', *Journal of Material Culture* 16(3): 241–63. https://doi.org/10.1177/1359183511412879.

Vilches, Flora, et al. 2014. 'La Minería de la Sal Durante el Siglo XX en San Pedro de Atacama, Chile (II Region): Entre la Explotación Artesanal y la Industrialización', *Estudios Atacameños* 48: 209–28. https://doi.org/10.4067/S0718-10432014000200014.

Villegas, Omar. 1987. *Atacameños: Una Nacion que Resiste*. Chile: Collection of Audiovisualist. Retrieved 31 July 2021 from https://www.facebook.com/omar.villegasastudillo.

Voskoboynik, Daniel Macmillen, and Diego Andreucci. 2021. 'Greening Extractivism: Environmental Discourses and Resource Governance in the 'Lithium Triangle'', *Nature and Space E* 5(2): 787–809. https://doi.org/10.1177/25148486211006345.

Wacquant, Loïc. 2012. 'Three Steps to a Historical Anthropology of Actually Existing Neoliberalism', *Social Anthropology* 20: 66–79. https://doi.org/10.1111/j.1469-8676.2011.00189.x.

Walsh, Casey. 2022. 'Beyond Rules and Norms: Heterogeneity, Ubiquity, and Visibility of Groundwaters', *WIREs Water* 9(4): e1597. https://doi.org/https://doi.org/10.1002/wat2.1597.

Walsh, Sarah. 2019. 'The Chilean Exception: Racial Homogeneity, Mestizaje and Eugenic Nationalism', *Journal of Iberian and Latin American Studies* 25 (1): 105–25. https://doi.org/10.1080/14701847.2019.1579499.

Weinberg, Marina. 2021. 'Cuerpos de Cobre: Extractivismo en Chuquicamata, Chile', *The Journal of Latin American and Caribbean Anthropology* 26(2): 200–18. https://doi.org/https://doi.org/10.1111/jlca.12545.

Welker, Marina. 2014. *Enacting the Corporation: An American Mining Firm in Post-Authoritarian Indonesia*. Berkeley: University of California Press.

West, Paige. 2016. *Dispossession and the Environment: Rhetoric and Inequality in papua New Guinea*. New York: Columbia University Press.

Weszkalnys, Gisa. 2011. 'Cursed Resources, or Articulations of Economic Theory in the Gulf of Guinea', *Economy and Society* 40(3): 345–72. https://doi.org/10.1080/03085147.2011.580177.

Whyte, Jessica. 2019. *The Morals of the Market: Human Rights and the Rise of Neoliberalism*. Brooklyn, NY: Verso.

Willow, Anna J. 2019. *Understanding ExtrACTIVISM: Culture and Power in Natural Resource Disputes*. New York: Routledge.

Winn, Peter. (ed.). 2004. *Victims of the Chilean Miracle: Workers and Neoliberalism in the Pinochet era, 1973–2002*. Durham: Duke University Press.

Wolf, Eric R. 1957. 'Closed Corporate Peasant Communities in Mesoamerica and Central Java', *Southwestern Journal of Anthropology* 13(1): 1–18. https://doi.org/www.jstor.org/stable/3629154.

———. 1986. 'The Vicissitudes of the Closed Corporate Peasant Community', *American Ethnologist* 13(2): 325–29. https://doi.org/www.jstor.org/stable/644135.

Yacktamarí. 2007. 'Revista de Arte y Cultura de Lickanckoi', *Periodico Lickanckoi*. Archive copy.

Yakovleva, Natalia. 2005. *Coporate Social Repsonsibility in the Mining Industries*. Burlington: Ashgate.

Yáñez, Nancy, and Raúl Molina. 2008. *La Gran Minería y los Derechos Indígenas en el norte de Chile*. Santiago: Observatorio de Derechos de los Pueblos Indígenas.

———. 2011. *Las Aguas Indígenas en Chile*. Santiago: LOM Ediciones.

INDEX

www.ingramcontent.com/pod-product-compliance
Ingram Content Group UK Ltd.
Pitfield, Milton Keynes, MK11 3LW, UK
UKHW021837150926
8656IPUK00004B/26